NOTICE

This document is disseminated under the sponsorship of the U.S. Department of Transportation in the interest of information exchange. The U.S. Government assumes no liability for its contents or use thereof. This report does not constitute a standard, specification, or regulation. The U.S. Government does not endorse products or manufacturers. Trade and manufacturers' names appear in this report only because they are considered essential to the object of the document.

Quality Assurance Statement

The Federal Highway Administration provides high quality information to serve Government, industry, and the public in a manner that promotes public understanding. Standards and policies are used to ensure and maximize the quality, objectivity, utility, and integrity of its information. FHWA periodically reviews quality issues and adjusts its programs and processes to ensure continuous quality improvement.

I0393874

1. Report No. FHWA-HEP-08-019	2. Government Accession No.	3. Recipient's Catalog No.
4. Title and Subtitle SAFETEA-LU 1808: Congestion Mitigation and Air Quality Improvement Program Evalutation and Assessment - Phase 1 Final Report		5. Report Date October 2008
		6. Performing Organization Code
7. Author(s) Michael Grant, Rich Kuzmyak, Lilly Shoup, Eva Hsu, Teddy Krolik, and David Ernst		8. Performing Organization Report No.
9. Performing Organization Name and Address ICF International 9300 Lee Highway Fairfax, Virginia 22031		10. Work Unit No. (TRAIS)
		11. Contract or Grant No. DTFH61-04-D-00015
12. Sponsoring Agency Name and Address Office of Natural and Human Environment Federal Highway Administration 1200 New Jersey Ave, SE Washington, DC 20590		13. Type of Report and Period Covered CMAQ Evaluation and review 2000 - 2005
		14. Sponsoring Agency Code HEPN-1

15. Supplementary Notes
This report was overseen by a review panel composed of representatives of Federal Highway Administration, Federal Transit Administration, and the U.S. Environmental Protection Agency.

16. Abstract
In SAFETEA-LU Section 1808, Congress required the U.S. Department of Transportation, in consultation with the Environmental Protection Agency (EPA), to evaluate and assess the direct and indirect impacts of a representative sample of Congestion Mitigation and Air Quality (CMAQ)-funded projects on air quality and congestion levels. This study responds to that request by analyzing 67 CMAQ-funded projects, using data supplied by States and Metropolitan Planning Organizations (MPOs) in the Federal Highway Administration (FHWA) CMAQ database. From this information, the study team examined the estimated impacts of these projects on emissions of transportation-related pollutants, including carbon monoxide (CO), ozone precursors – oxides of nitrogen (NOx) and volatile organic compounds (VOCs) – and particulate matter (PM10 and PM2.5), as well as on traffic congestion and mobility. The study team also conducted additional analyses of the selected set of CMAQ-funded projects to estimate their cost-effectiveness at reducing emissions of each pollutant.

17. Key Words The Congestion Mitigation and Air Quality Improvement Program (CMAQ), CMAQ, air quality, evaluation, assessment, cost-effectiveness, SAFETEA-LU 1808		18. Distribution Statement No restrictions. This document is available to the public electronically through the Federal Highway Administration Office of Natural and Human Environment, Washington DC, 20590 http://www.fhwa.dot.gov/environment/cmaqpgs/index.htm.	
19. Security Classif. (of this report) Unclassified	20. Security Classif. (of this page) Unclassified	21. No. of Pages 152	22. Price

Form DOT F 1700.7 (8-72)   Reproduction of completed page authorized

Contents

List of Figures

List of Tables

EXECUTIVE SUMMARY

The Congestion Mitigation and Air Quality Improvement (CMAQ) Program provides funds to States for transportation projects designed to improve air quality and reduce traffic congestion, particularly in areas of the country that do not attain national air quality standards. Created by the Intermodal Surface Transportation Efficiency Act (ISTEA) of 1991, the program was reauthorized under the Transportation Equity Act for the 21st Century (TEA-21) in 1997, and again as part of the Safe, Accountable, Flexible, Efficient, Transportation Equity Act: A Legacy for Users (SAFETEA-LU) in 2005. Since 1991, the CMAQ Program has provided funding to over 16,000 projects, and has been a key mechanism for supporting investments that help urban areas meet air quality goals, encourage alternatives to driving alone, and improve traffic flow.

In SAFETEA-LU Section 1808, Congress required the U.S. Department of Transportation, in consultation with the Environmental Protection Agency (EPA), to evaluate and assess the direct and indirect impacts of CMAQ-funded projects on air quality and congestion levels. This study responds to that request by analyzing 67 CMAQ-funded projects, using data supplied by States and metropolitan planning organizations (MPOs) in the Federal Highway Administration (FHWA) CMAQ database. From this information, the study team examined the estimated impacts of these projects on emissions of transportation-related pollutants, including carbon monoxide (CO), ozone precursors – oxides of nitrogen (NOx) and volatile organic compounds (VOCs) – and particulate matter (PM_{10} and $PM_{2.5}$), as well as on traffic congestion and mobility. The study team also conducted additional analyses of the selected set of CMAQ-funded projects to estimate their cost-effectiveness at reducing emissions of each pollutant.

Congestion and Mobility Benefits

As shown in the set of projects examined in this study, many CMAQ projects help to reduce traffic congestion and improve mobility. Traffic flow improvement projects, which include traffic signalization improvements, incident management programs, and intersection improvements, are designed to improve traffic speeds and minimize delays experienced by drivers. Although many of these projects are small in scope (e.g., an individual intersection improvement), they can have a sizable impact on travel times in specific locations. For instance, among the traffic flow projects that reported travel time savings – installation of coordinated signalized intersections along a roadway in Newark, Ohio; two intersection improvements in East Baton Rouge Parish, Louisiana; and traffic signal optimization for arterial highways in Lexington, Kentucky – the projects were estimated to save from 702 to 6,360 vehicle hours of delay per day. In total, traffic flow improvement projects represented 42 percent of the CMAQ-funded projects (32 percent of CMAQ funding) during fiscal years (FY) 2000 to 2005, according to the FHWA CMAQ database. In addition to traffic flow improvements, freight and intermodal projects are frequently designed to shift goods movement from trucks to rail, and reduce congestion associated with truck traffic on major freight corridors.

Other projects are designed primarily to enhance mobility by increasing travel options, such as transit, bicycling, walking, and ridesharing. Most of the vanpool, park-and-ride, bicycle/pedestrian, and transit bus service improvement projects examined in this study were estimated to remove from about one hundred to several hundred vehicle trips per day. The magnitude of congestion relief effects from this level of vehicle travel reduction is difficult to assess, and was typically not reported by project sponsors. The primary benefit from these projects is enhanced mobility for travelers, as travelers have a greater range of options to meet their travel needs and have greater access to employment, services, and recreational opportunities. These projects often also have other benefits such as reducing travel costs for individuals and supporting improved quality of life in communities. Mobility can also be enhanced through projects that improve incident management, freeway traveler information, and transit information, which improve travel time reliability and enable people to plan their travel routes, mode choice, and time of travel more effectively.

Air Quality Benefits

CMAQ projects typically reduce motor vehicle emissions one of three ways: 1) by encouraging changes in travel behavior that reduce motor vehicle miles traveled (VMT), such as shifts to ridesharing, transit, bicycling, or walking; 2) by improving traffic flow, thereby reducing vehicle idling and stop-and-start driving conditions that are associated with higher levels of emissions; and 3) by implementing technologies to reduce the rate of emissions, such as conversion to alternative fueled buses, or retrofits of diesel vehicles. In addition, in some locations, targeted approaches have been used to reduce wind blown particulate matter from roadways, such as funding street sweepers and application of de-icing chemicals instead of sand.

Although the limited number of projects examined in this study does not allow for definitive conclusions about the effectiveness or cost-effectiveness of different types of CMAQ projects, some general findings are noted below.

First, since many CMAQ projects are small in scale (e.g., a single park-and-ride lot, a bicycle path, a transit shuttle), many of these projects yield small reductions in motor vehicle emissions. Among the projects reviewed in this study, the majority had emissions reduction estimates of less than 50 kg per day of both VOC and NO_X, and less than 500 kg per day of CO. In the context of regional air quality concerns, these estimated emissions reductions are generally quite small. The combined impact of multiple projects, and longer-term, indirect benefits (e.g., supporting transit-oriented land use patterns), however, may be more substantial.

Second, a wide variation in estimated emissions effects and cost-effectiveness occurs within project types. Some individual projects showed very strong cost-effectiveness, while other similar types of projects appeared to have poor cost-effectiveness at reducing specific pollutants. Within a given project category, estimated project cost-effectiveness typically varied by a factor of 10 or more (e.g., the most cost-effective new bus service in the set of projects examined was estimated to cost $130,000/ton of VOC removed, while the least cost-effective new bus service was estimated to cost $1.5 million/ton of VOC removed). This high level of variability suggests that local context and project-specific factors are important determinants of the level of emissions reductions that can be expected from projects.

Third, although there is a wide range of estimated emissions benefits and cost-effectiveness at reducing emissions across the set of projects examined, there are some patterns when looking at impacts on individual pollutants. Strategies that aim to reduce vehicle travel, such as shared ride programs, travel demand management, bicycle/pedestrian facilities, and transit improvements, typically reduce emissions of all major on-road transportation related pollutants – VOC, NO_X, CO, and PM_{10} and $PM_{2.5}$ – with the largest reductions occurring in ozone precursors and CO. PM reductions from these projects tended to be very small and in many cases were not reported by project sponsors.

Traffic flow improvements, such as signal syncronization and freeway management projects, are typically implemented to improve travel speeds on congested roadways, or to reduce idling time. The emissions effects of traffic flow improvements depend on the overall speed improvement and initial speeds. VOC emissions generally decline with increasing speeds, but NO_X and CO emissions can increase at higher speeds. As a result, a traffic flow project could reduce VOC emissions but yield a small increase in NO_X, and may have little or no effect on PM.

Finally, diesel emissions-focused strategies can be quite cost-effective at reducing PM emissions. Among the sample projects, dust mitigation-focused projects offered the most cost-effective means for reducing PM_{10} and $PM_{2.5}$ from wind-blown dust in locations where they were practical. Diesel engine retrofits and truck idle reduction strategies tended to be the most cost-effective set of strategies for reducing particulate matter outside of the dust mitigation strategies. This is perhaps not surprising, given that diesel vehicles are large emitters of particulate matter, but it is also notable that some diesel engine retrofit projects examined in this study were quite cost-effective at reducing ozone precursors and CO as well. For instance, one type of diesel soot filter used to retrofit transit buses was certified to reduce PM, VOC, and CO emissions each by 60 percent; another technology used in a project to retrofit trash collection trucks was estimated to reduce PM emissions by 80 percent, while also reducing CO by 67 percent and VOC by 95 percent.

Effective Implementation of the CMAQ Program

In addition to determining the impacts of a sample of CMAQ projects on air quality and congestion, SAFETEA-LU Section 1808 directs an evaluation and assessment of CMAQ projects to "ensure the effective implementation of the program." This report is the first phase of a two phase effort being undertaken by DOT, in consultation with EPA, to address the goals of this evaluation and assessment. This Phase I report focuses on an evaluation of a set of CMAQ projects for the purpose of determining their air quality and congestion benefits, while Phase II involves case studies to further explore approaches to CMAQ project selection and implementation that are effective in achieving air quality improvement and congestion relief.

In the course of collecting data on the selected projects a variety of good practices that States and MPOs use to analyze and select projects for CMAQ funding were revealed. These approaches include: development of standardized templates, calculation guidebooks, and spreadsheets that help to ensure a consistent set of project inputs from project sponsors and to make calculations easier and less prone to error; development of systematic procedures for ranking projects, including consideration of project cost-effectiveness at reducing air pollutant emissions of concern and other factors; and coordination with air agencies and local agencies in the project selection process. The information gathered for this Phase I report was used to help select locations for case study visits in Phase II.

The analysis of emissions reduction cost-effectiveness in this study also provides a possible analytic framework that may help States and MPOs develop their own analysis when considering projects for funding. It is important to note, however, that CMAQ projects also generate other benefits beyond emissions reductions, such as congestion relief, travel time savings, energy savings, enhanced mobility, and other transportation system user benefits, which are not quantified in the emissions reduction cost-effectiveness figures but are important considerations in the overall benefit-cost associated with each project. These many factors also are often important considerations in project selection. Many States and MPOs value the CMAQ Program for the flexible funding it provides to help them address air quality concerns from transportation sources and to help support a wide range of transportation objectives, such as enhancing multi-modal accessibility, improving transportation system reliability, and strengthening community livability.

1.   INTRODUCTION

Purpose of the Study

The Congestion Mitigation and Air Quality Improvement (CMAQ) Program provides funds to States for transportation projects designed to reduce traffic congestion and improve air quality, particularly in areas of the country that do not attain national air quality standards. Created by the Intermodal Surface Transportation Efficiency Act (ISTEA) of 1991, the program was reauthorized under the Transportation Equity Act for the 21st Century (TEA-21) in 1998 and again as part of the Safe, Accountable, Flexible, Efficient, Transportation Equity Act: A Legacy for Users (SAFETEA-LU) in 2005. From its beginning, the CMAQ program has been a key funding mechanism for helping urban areas meet air quality goals and supporting investments that encourage alternatives to driving alone and improve traffic flow. Since 1991, the Program has provided $22.7 billion in funding to States, metropolitan planning organizations (MPOs), and transit agencies to invest in projects that reduce criteria air pollutants regulated from transportation-related sources. The CMAQ program is also credited with gaining State and regional support for the mandates of the 1990 Clean Air Act Amendments (CAAA).[1]

In TEA-21, Congress funded a study to better understand the efficiency and effectiveness of the CMAQ program, which was undertaken by the Transportation Research Board (TRB). Building on this effort, the further reauthorization of the CMAQ program in Section 1808 of SAFETEA-LU requires the U.S. Department of Transportation, in consultation with the U.S. Environmental Protection Agency (EPA), to conduct an evaluation and assessment of a representative sample of CMAQ projects, for the purpose of: (A) determining their congestion and air quality benefits, and (B) ensuring effective implementation of the program. Moreover, SAFETEA-LU placed increased emphasis on funding cost-effective strategies, calling for States and MPOs to give priority in distributing funds to diesel retrofits and "cost-effective congestion mitigation activities that provide air quality benefits."[2]

The language of SAFETEA-LU Section 1808 requiring an evaluation and assessment of CMAQ projects is included below.

> (f) EVALUATION AND ASSESSMENT OF CMAQ PROJECTS.—Section 149 of such title (as amended by subsection (e)) is amended by adding at the end the following:
>
> (h) EVALUATION AND ASSESSMENT OF PROJECTS.—
>
> (1) IN GENERAL.—The Secretary, in consultation with the Administrator of the Environmental Protection Agency, shall evaluate and assess a representative sample of projects funded under the congestion mitigation and air quality program to—
>
> (A) determine the direct and indirect impact of the projects on air quality and congestion levels; and
>
> (B) ensure the effective implementation of the program.
>
> (2) DATABASE.—Using appropriate assessments of projects funded under the congestion mitigation and air quality program and results from other research, the Secretary shall maintain and disseminate a cumulative database describing the impacts of the projects.
>
> (3) CONSIDERATION.—The Secretary, in consultation with the Administrator, shall consider the recommendations and findings of the report submitted to Congress under section 1110(e) of the Transportation Equity Act for the 21st Century (112 Stat. 144), including recommendations and findings that would improve the operation and evaluation of the congestion mitigation and air quality improvement program.

This report is the first phase of a two phase effort being undertaken by FHWA, in consultation with EPA, in order to meet the requirements in Section 1808(f) of SAFETEA-LU. The purpose of this report is to

---

[1] Transportation Research Board. *Special Report 264: The CMAQ Program: Assessing 10 Years of Experience.* 2002. Page 19.

[2] SAFETEA-LU 1808(d) amending 23 USC 149 (f)(3)(A)(ii).

examine the direct and indirect impacts of CMAQ-funded projects on air quality and congestion levels. This evaluation was conducted by gathering data reported in FHWA's national database of CMAQ projects, as well as additional background collected from States and MPOs to analyze the total annual costs (i.e., CMAQ and non-CMAQ funds), estimated annual emissions reductions, and congestion relief benefits for a small number of CMAQ funded projects. The report also contains an assessment of the air quality cost-effectiveness of these selected projects, and preliminary information on good practices being implemented by State DOTs and MPOs for prioritizing and selecting CMAQ projects. This preliminary information is followed by a Phase II report that involves case studies of a sample of State DOTs and MPOs to highlight approaches that advance the effective implementation of the program.

Context for the CMAQ Program

Any evaluation of CMAQ projects should recognize the magnitude of the air quality and congestion problems in the United States and have realistic expectations concerning the influence one program can have on reducing transportation-generated pollution and mitigating traffic congestion. Despite substantial progress in improving air quality nationally since the 1970s, over 100 million Americans still live in areas of the country that do not meet EPA's National Ambient Air Quality Standards for one or more pollutants. Traffic congestion is a growing problem affecting urban areas of all sizes, with congestion affecting more trips, over more hours of the day, on more roadways than in the past. According to Texas Transportation Institute's *Urban Mobility Study*, traffic congestion creates a $78 billion annual drain on the U.S. economy in the form of 4.2 billion lost hours and 2.9 billion gallons of wasted fuel.[3]

While the CMAQ program provides targeted resources to address the role of transportation in these air quality and congestion challenges, the resources provided by the CMAQ program are modest in comparison to the overall Federal transportation program. In total, SAFETEA-LU provides $286.4 billion in guaranteed funding for Federal surface transportation programs over five years through FY 2009. This includes $193.6 billion in Federal-aid Highway program authorizations and $52.6 billion for Federal transit programs, as well as other projects. The CMAQ program is authorized at $8.6 billion, or 4.4 percent of the total Federal-aid Highway program (three percent of the total Federal surface transportation program funding). Given other State and local sources of funding, which make up about half of all highway and transit capital expenditure and the majority of operating expenses, the Federal CMAQ program represents less than two percent of total transportation spending in many metropolitan areas.

A single major transportation infrastructure project in an urban area can cost more than $1 billion, and there are a number of major highway and transit projects being constructed across the U.S. that cost in the multiple billions of dollars. At an authorized level of approximately $1.7 billion per year under SAFETEA-LU, the CMAQ program – which provides funding to all 50 States and the District of Columbia – is not able to substantially "solve" the air quality or congestion problems facing metropolitan areas across the country. However, the incremental benefits of the program are an important part of the solution.

The CMAQ program provides funds that are targeted to areas of the country with the most severe air quality problems, which tend to be the largest metropolitan areas experiencing some of the worst traffic congestion. Many metropolitan areas rely on the CMAQ program as a flexible funding source to support a wide range of projects that improve air quality, reduce traffic congestion, and support a multi-modal transportation system, and as a mechanism to help fund air quality mandates under the Clean Air Act. CMAQ funded projects are often small in scale – e.g., a bicycle path, a park-and-ride lot, a new transit shuttle service, or a traffic signalization improvement. Still, they may have important benefits at a corridor or local level, where the benefits of a single project can make a difference. CMAQ funds also are

---

[3] Texas Transportation Institute, *2007 Urban Mobility Report.* September 2007. Available at: http://mobility.tamu.edu/ums/report/.

used to leverage other Federal and State and local funding sources, and to support regional efforts such as regional ridesharing programs, incident management programs, and traveler information systems.

Establishment of the CMAQ Program

The Intermodal Surface Transportation Efficiency Act (ISTEA) of 1991 was a landmark surface transportation Act which, for the first time, emphasized intermodalism - the seamless linking of highway, rail, bicycle, pedestrian, and other modes. The Act included provisions and new programs designed to address the Nation's growing transportation challenges, such as improving safety, reducing traffic congestion, improving efficiency in freight movement, increasing intermodal connectivity, and protecting the environment. ISTEA opened the transportation planning process to more public involvement than ever before, bringing new players to the table when decisions were being made and increasing collaboration among existing stakeholders. This diversity in transportation decision-making has resulted in additional positive benefits for communities, because transportation investment decisions are built upon input from transportation stakeholders and the general public.

ISTEA also made funding available to new kinds of programs, and established the Congestion Mitigation and Air Quality Improvement (CMAQ) program – the first federally funded transportation program explicitly targeting air quality improvement. Approximately 4 percent of total funding for the 1992–1997 Federal surface transportation program, or $6 billion, was authorized for CMAQ projects that would offer alternatives to single occupant vehicle (SOV) travel, improve travel efficiency as a means of addressing traffic congestion, and promote cleaner motor vehicles in the Nation's most polluted areas.[4]

From its inception, the primary policy focus of the CMAQ program has been on air quality improvement, reflecting the requirements placed on the transportation sector by the Clean Air Act Amendments (CAAA) of 1990 to help meet national air quality goals. It provides flexible funding for States to use in nonattainment areas to help them address air quality concerns from transportation sources. Over time, the CMAQ program has become a key funding mechanism to support investments that not only help urban areas meet air quality goals, but also help focus transportation planning on a more inclusive, environmentally sensitive, and multimodal approach.

Apportionment and Eligible Projects

Federal CMAQ funds are apportioned annually to each State according to the severity of the air quality problem and the population of each nonattainment or maintenance county (based upon Census Bureau data).[5] Each State is guaranteed a minimum apportionment of one-half percent of the year's total program funding, regardless of whether the State has any nonattainment or maintenance areas. These minimum apportionment funds can be used anywhere in the State for projects eligible for either CMAQ or the Surface Transportation Program (STP).

To be eligible for CMAQ funds, a project must be included in the MPO's current transportation plan and Transportation Improvement Program (TIP) (or the current Statewide TIP, or STIP in areas without an MPO).[6] In nonattainment and maintenance areas, the project also must meet the conformity provisions

---

[4] See Transportation Research Board. *Special Report 264: The CMAQ Program: Assessing 10 Years of Experience.* 2002. Page 19.

[5] 23 USC 149(b)-(c).

[6] 23 USC 134-35, 149(d).

contained in Section 176(c) of the Clean Air Act and the transportation conformity rule at 40 CFR Part 93. In general, there are three types of CMAQ eligible activities:[7]

- Capital Investment - CMAQ funds may be used to establish new or expanded transportation projects or programs that reduce emissions, including capital investments in transportation infrastructure, congestion relief efforts, diesel engine retrofits, or other capital projects.

- Operating Assistance - Operating assistance is limited to new transit services, intermodal facilities, and travel demand management strategies (including traffic operation centers); and the incremental cost of expanding existing transit services. In using CMAQ funds for operating assistance, the intent is to help start up viable new transportation services that can demonstrate air quality benefits and eventually cover their costs as much as possible. Once these projects have become part of the baseline transportation network and no longer represent additional air quality benefits, other funding sources should supplement and ultimately replace their CMAQ funds for operating assistance (i.e., there is a three-year limit on the use of CMAQ operating assistance).

- Planning and Project Development - Activities in support of eligible projects also may be appropriate for CMAQ investments. Studies that are part of the project development pipeline (e.g., preliminary engineering) under the National Environmental Policy Act (NEPA) are eligible for CMAQ support, as are FTA's Alternatives Analyses. General studies that fall outside specific project development do not qualify for CMAQ funding. Examples of such efforts include major investment studies, commuter preference studies, modal market polls or surveys, transit master plans, and others. These activities are eligible for Federal planning funds.

The CMAQ *Interim Program Guidance* dated October 31, 2006 lists 16 categories of projects eligible for CMAQ funding, which FHWA has traditionally grouped into the following categories:

- Traffic flow improvements (e.g., traffic signalization, freeway management, high-occupancy vehicle lanes);

- Shared ride programs (e.g., regional ridesharing, vanpool programs, and park-and-ride lots);

- Travel demand management (e.g., regional marketing, employer trip reduction programs);

- Bicycle/pedestrian facilities and programs;

- Transit (e.g., new bus services, new rail services/equipment, service upgrades/amenities, bus replacements, alternative fuel buses); and

- Other projects, including diesel engine retrofits, freight/intermodal projects, dust mitigation projects, and other qualifying projects, including experimental pilot projects, which are allowed under the law as demonstrations to determine their benefits and costs.

CMAQ-funded projects or programs must reduce CO, ozone precursor (NOx and VOCs), PM, or PM precursor (e.g., NOx) emissions from transportation.[8] These reductions should contribute to the area's

---

[7] FHWA Memorandum. October 31, 2006. "Guidance on the Congestion Mitigation and Air Quality Improvement (CMAQ) Program Under the Safe, Accountable, Flexible, Efficient Transportation Equity Act: A Legacy for Users." Pages 10 – 11, interpreting 23 USC 149(b).

[8] 23 USC 149(b).

overall clean air strategy.[9] The traditional Federal share for most eligible CMAQ projects has been 80 percent. However, some projects that also focus on safety efforts have been funded at 100 percent Federal share.[10] More recently, the 2007 Energy Independence and Security Act provides the option of 100 percent Federal share for all CMAQ projects in FY 2008 and FY 2009.[11]

State DOTs and MPOs, the key agencies for transportation planning at the local and regional level, operate under broad guidance regarding project eligibility when they determine project selection and implementation decisions. Notwithstanding the statutory formula for determining the apportionment amount, the State may use its CMAQ funds in any ozone, CO, or PM nonattainment or maintenance area. States may do so according to local preference; there is no obligation to allocate CMAQ funds in the same way they are apportioned.

CMAQ projects are usually proposed and evaluated at the MPO level and then chosen at the State level using a variety of selection processes. To support a transparent, open process, MPOs, State DOTs, and transit agencies are encouraged to establish and make available a project selection process that clearly identifies the basis for selecting projects, including emissions benefits, cost effectiveness, and additional selection factors such as congestion relief, greenhouse gas reductions, safety, system preservation, access to opportunity, sustainable development and freight, reduced SOV reliance, multi-modal benefits, or other criteria. At a minimum, projects should be identified by year and proposed funding source.[12]

Required Emissions Analyses

The CMAQ statute includes emissions reduction as a requirement for CMAQ-invested projects or programs.[13] Project sponsors must estimate the expected emissions reductions for projects funded under the program, with particular attention to the pollutants of concern in the project implementation area (CO, VOCs, NOx, $PM_{2.5}$ and PM_{10}).[14] According to the Interim Program Guidance, quantified emissions benefits (i.e., emissions reductions) and emissions increases should be included in all project proposals, except where it is not possible to quantify emissions changes, in which case a qualitative assessment may be provided. Emissions effects should be estimated and reported in a consistent fashion (i.e., kg/day) across projects to allow accurate comparison during the project selection process.

State and local transportation and air quality agencies may conduct CMAQ-project air quality analyses with different approaches; FHWA does not specify the emissions reduction methodologies to be used. However, FHWA stipulates that every effort should be taken to ensure that determinations of air quality benefits are credible and based on a reproducible and logical analytical procedure for inclusion in FHWA's national CMAQ database.[15]

---

[9] FHWA Memorandum. October 31, 2006. "Guidance on the Congestion Mitigation and Air Quality Improvement (CMAQ) Program Under the Safe, Accountable, Flexible, Efficient Transportation Equity Act: A Legacy for Users." Page 11, interpreting 23 USC 149(b).

[10] 23 USC 120(c).

[11] See Section 1131 of the Energy Independence and Security Act of 2007 (P.L. 110-140, H.R. 6).

[12] See FHWA Memorandum. October 31, 2006. "Guidance on the Congestion Mitigation and Air Quality Improvement (CMAQ) Program Under the Safe, Accountable, Flexible, Efficient Transportation Equity Act: A Legacy for Users." Page 26, interpreting 23 USC 149(b).

[13] The exception is for states receiving minimum apportionment that do not have, and have had, a nonattainment area designed under the Clean Air Act, in which case the State may use funds for any project eligible under the Surface Transportation Program. 23 USC 149(b)-(c).

[14] 23 USC 149(b)-(c).

[15] See FHWA Memorandum. October 31, 2006. "Guidance on the Congestion Mitigation and Air Quality Improvement (CMAQ) Program Under the Safe, Accountable, Flexible, Efficient Transportation Equity Act: A Legacy for Users." Page 25.

Relevant Literature and Studies

While this study focuses on a set of CMAQ-funded projects from FHWA's national CMAQ database, it also builds upon and is intended to supplement past efforts to assess the impacts of CMAQ projects.

Since the program's inception, FHWA and EPA have developed several documents that have described the emissions benefits, congestion benefits, and other positive effects of CMAQ funded projects. For instance, FHWA developed a document in 1996, *Congestion Mitigation and Air Quality Improvement Program – Indirect Benefits*, which highlights the CMAQ program's indirect benefits to MPOs and other stakeholders in the transportation planning process, and provided examples of specific projects and their benefits.[16] In 1999, EPA created a brochure on the CMAQ program, "Creating Transportation Choices: Congestion Mitigation and Air Quality Improvement Program Success Studies," which highlights a set of examples of CMAQ-funded projects.[17] In 2003, FHWA developed a report, *CMAQ: Advancing Mobility and Air Quality,* which describes the ways in which CMAQ projects can improve mobility.[18] The report documents nine examples of CMAQ projects and how they have enhanced mobility, and includes information on the emissions benefits reported for these projects.

For over a decade, FHWA and EPA also have undertaken efforts to assess the effectiveness of the CMAQ program, to examine cost-effectiveness, and to provide information on recommended practices for estimating emissions effects. In 1997, FHWA sponsored a literature review on the *Cost Effectiveness of Transportation Control Measures by CMAQ Category*, which documented a range of studies summarizing the emissions benefits of projects funded under CMAQ.[19] To address concerns about the effectiveness of the CMAQ program at reducing motor vehicle emissions during the deliberations that led to passage of TEA-21, the EPA in coordination with FHWA, conducted a detailed assessment of CMAQ project effects and costs (*Costs and Emissions Benefits of CMAQ Project Types*, 1999). The study documented emissions effects, costs, methodologies and assumptions for 24 CMAQ projects within six project categories.[20]

In TEA-21, Congress called for an evaluation of the benefits and cost-effectiveness of projects funded under the CMAQ program. The Transportation Research Board (TRB) conducted this study, which examined the emissions benefits of CMAQ-funded projects, based on available literature and conducted a comparison against other pollution reduction measures to evaluate the program's cost effectiveness. Overall, the TRB study (published as Special Report 264) concluded that the CMAQ program had been valuable to its designed objectives, and supported its continuation subject to some targeted findings and recommendations. Perhaps the most critical finding in the review was the inability to evaluate performance of funded projects due to poor or absent information. This resulted in a recommendation for a more formal evaluation component attached to future funding of the CMAQ program. The study noted that the CMAQ program had not been structured to be evaluated in a rigorous way (local flexibility was an important feature of the program), thus making it impossible to perform a rigorous scientific analysis of benefits of CMAQ-funded projects.[21]

---

[16] Federal Highway Administration, *Congestion Mitigation and Air Quality Improvement Program: Indirect Benefits*. Publication No. FHWA-PD-97-045, 1997.

[17] Hagler Bailly, "Creating Transportation Choices: Congestion Mitigation and Air Quality Improvement Program Success Studies," for U.S. EPA, 1999.

[18] Federal Highway Administration, *CMAQ Advancing Mobility and Air Quality*. Publication No. FHWA-EP-03-045, 2003.

[19] Center for Transportation and the Environment (CTE), North Carolina State University, *Cost Effectiveness of Transportation Control Measures by CMAQ Category*, FHWA, 1997.

[20] Hagler Bailly, *Costs and Emissions Impacts of CMAQ Project Types*, U.S. EPA, 1999.

[21] Transportation Research Board. *Special Report 264: The CMAQ Program: Assessing 10 Years of Experience*. 2002.

This study is designed to supplement the findings of previous research by examining a small number of CMAQ-funded projects, with a focus on projects funded in years 2000 and later. By examining CMAQ projects that have been implemented, this study provides information on the estimated impacts that have been achieved on congestion levels and air quality.

Report Organization

This report is organized into three major sections:

- **Study Approach (Section 2)** discusses the parameters and methodology used for this study, including the approach and methodology used for gathering project information.

- **Impacts of Projects on Air Quality and Congestion (Section 3)** presents the results of the review of 67 projects. It reports on the estimated congestion benefits, emissions benefits, and costs of the selected projects, organized by project category for analysis purposes. This analysis relies on data provided by the project sponsors, used in their own analyses for reporting to FHWA's CMAQ database.

- **Project Analysis and Selection Practices that Support Effectiveness (Section 4)** uses information from the set of projects to assess cost-effectiveness at reducing emissions, and examines how State and local agencies are using this type of information for project prioritization and decision making. First, it examines the cost-effectiveness of the CMAQ projects by project category in dollars per ton of pollution reduced. For this analysis, emissions estimates of the projects were generally recalculated using standardized emissions factors (rates of emissions in grams per mile) in order to fill in gaps in reported emissions effects and to enable comparisons among projects that were implemented in different locations at different times. This section then includes a preliminary discussion of some approaches that have been used by States and MPOs to enhance the effectiveness of their project selection processes, drawing on the data collection effort conducted for this study. This preliminary information will form the basis for selecting locations for cases studies, which will be conducted in Phase II to examine program implementation at the State and local level.

Appendixes provide additional information on assumptions used in the calculations, and include short write-ups of each project in a standardized template.

2.   STUDY APPROACH

As noted previously, this evaluation responds to a desire by Congress to better understand the direct and indirect impact of CMAQ projects on air quality and congestion levels after more than 15 years of experience. Congress also wanted to ensure the effective implementation of the program, make sure the DOT maintain and disseminate a cumulative database of annual CMAQ reports and that the DOT and EPA consider recommendations to improve the operation and evaluation of the CMAQ program.

The study team determined it was best to approach the request in two phases. This Phase I report is intended to satisfy the understanding of the air quality and congestion benefits of a sample of CMAQ projects. To conduct the evaluation, a small set of CMAQ projects was chosen, background data were collected, and analysis was conducted to determine the effects of these projects on emissions and on congestion levels. In addition, the research team collected data on the costs of these projects – both from CMAQ funding and other funding sources – and conducted additional analyses to assess the cost-effectiveness of the sample projects at reducing emissions of each pollutant. In Phase II, FHWA, in consultation with EPA, conducted case studies of several States and metropolitan areas to understand project analysis, selection, and prioritization procedures.

This section describes the approach to selecting the CMAQ projects and collecting data used in the Phase I study.

Project Categories and Distribution

The research team attempted to select a set of projects for evaluation that would reflect typical projects funded through the program. As noted earlier, the CMAQ program has traditionally organized projects into several large categories. Figure 1 presents a breakdown of the number of CMAQ funded projects by these major categories for FY 2000 to FY 2005.

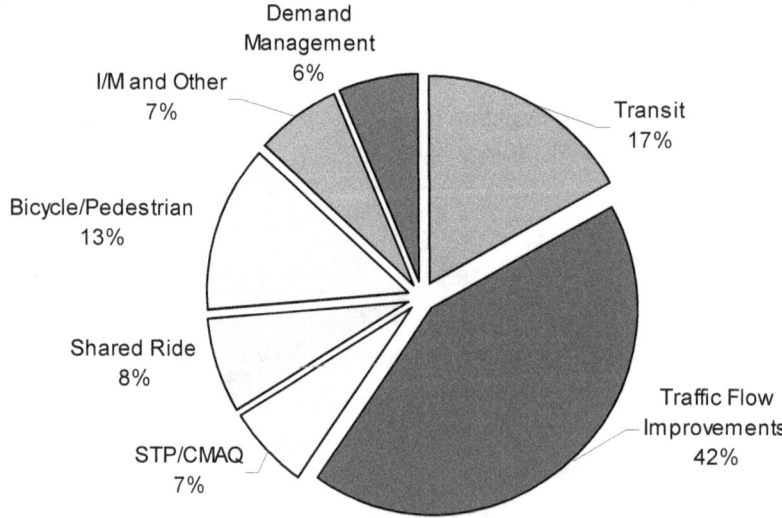

Figure 1. Share of CMAQ Projects Obligated FY 2000 to FY 2005 by Project Category.

Figure 2 presents a breakdown of the percentage of total CMAQ funding received by projects in these major categories for FY 2000 to FY 2005.

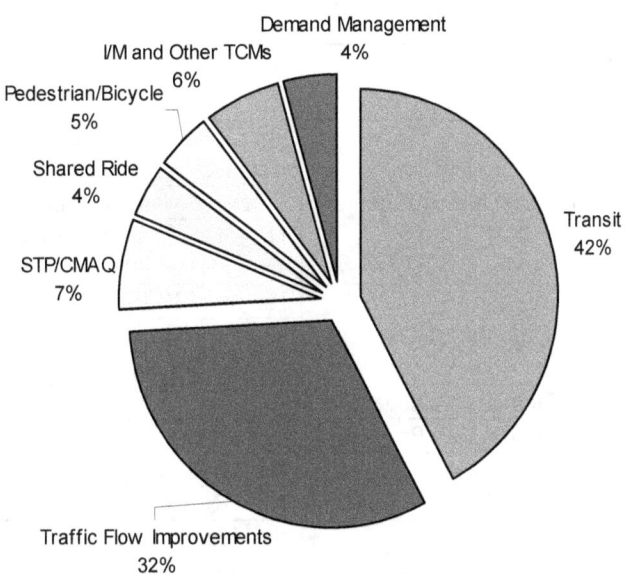

Figure 2. Percentage of CMAQ Funding FY 2000 to FY 2005 by Project Category.

As can be seen from these two diagrams, transit and traffic flow improvement projects together made up over half the total number of projects funded and received nearly three-quarters of total funding. Transit projects, in particular, made up a larger share of funding than share of projects because many transit projects were larger in scale and involved more capital funding (e.g., vehicle purchases, new transit lines, etc.). In contrast, pedestrian/bicycle projects made up a smaller share of funding than their proportion of total projects since these projects tended to be small and involved less funding per project.

In developing a set of projects for analysis in this study, the CMAQ projects were organized into the major categories that FHWA has traditionally used to group CMAQ projects in its database. Subcategories were broken out where categories are larger and diverse, and additional categories were created for project types that currently are the focus of additional attention (e.g., diesel engine retrofits, freight/intermodal projects).

Given that more than half the projects funded by CMAQ over this period were either traffic flow improvements or transit projects, the study team determined that in order to ensure an adequate number of projects across the various project categories, the number of projects analyzed in each category would not be proportional to the number of funded projects or total funding by category. Rather, to ensure a reasonable number of projects across all categories, the study team endeavored to include a minimum of three projects within each subcategory. Larger samples were included for categories that historically have had more CMAQ projects funded or that are currently the focus of additional attention. In two cases (High-Occupancy Vehicle Lanes and Conventional Bus Replacements), fewer projects were collected, due to data limitations and relatively few projects that have been funded in recent years for these types of projects.

Table 1 illustrates the project categories and subcategories, and the number of CMAQ projects that were analyzed.

Table 1. CMAQ Projects Included in the Study by Project Category and Subcategory.

Category	Subcategory	Number of Projects
Traffic Flow Improvements	Traffic Signalization/ Intersection Improvements	6
	Freeway Management	4
	High-Occupancy Vehicle Lanes	1
Shared Ride Programs	Regional Ridesharing	3
	Vanpool Programs	4
	Park and Ride lots	5
Travel Demand Management		4
Bicycle/Pedestrian Projects		4
Transit Service Improvements	New Bus Services	3
	New Rail Services	3
	Service Upgrades/Amenities	5
Transit Vehicle Replacements and Related Infrastructure	Conventional Bus Replacements	2
	Alternative Fuel Vehicles/Fueling Facilities	4
Dust Mitigation		3
Freight/Intermodal		6
Diesel Emissions Reduction	Diesel Engine Retrofits	7
	Truck Idle Reduction	3
	Total	*67*

Project Selection Procedures

Once the broad organization of projects was determined, a screening analysis based solely on information contained in the CMAQ database was conducted. The CMAQ database is a tool developed by FHWA to capture information about CMAQ projects, including funding information, emissions reduction estimates, the MPO, nonattainment or maintenance area, and a brief project description.

The process of selecting a small set of projects for analysis focused on identifying entries in the database which met the following selection criteria:

- Projects funded in the FY 2000 funding cycle or later;
- Quantitative emissions reductions were reported for at least one of the following pollutants: NO_X, VOC, CO, or PM;
- Reported emissions effects appeared "reasonable", based on judgment of the study team (i.e., eliminating projects with suspect or unusual emissions results); and
- Projects represented "typical" projects within the category (i.e., eliminating specialized or unusual project examples, based on project description).

In addition to the above selection criteria, the initial screening focused largely, but not exclusively, on States that are apportioned the highest levels of CMAQ funding. This was done to achieve a representative sample of projects, since these States fund a larger number of projects; also, staff within these States would be able to provide information on more projects per contact should follow-up be needed. (See Table 2 for a list of States with the largest CMAQ apportionments over the period FY 1991 to FY 2005, along with the amount obligated.) A final consideration was the overall geographic diversity of the sample, and balance between large projects and small projects. The CMAQ funds received by the selected projects ranged in scope from $33,000 for a signal synchronization project in Tennessee to $36.2 million for a transit improvement project that involved double tracking segments of a commuter rail line between Dallas and Ft. Worth. In some cases, CMAQ funded only a small portion of the total project costs, helping to leverage other Federal (specifically, FTA) or state and local funds; this was the case in particular for some of the larger transit projects and HOV project. In many other cases, CMAQ provided the majority or sole source of funding.

Table 2. States Receiving Largest CMAQ Apportionments, FY 1991 – FY 2005.

State	Amount Apportioned (Million $)	Amount Obligated (Million $)
California	$ 4,019.1	$ 3,637.8
New York	$ 1,698.6	$ 1,401.7
Texas	$ 1,469.8	$ 1,208.7
New Jersey	$ 1,084.2	$ 987.4
Illinois	$ 950.1	$ 817.1
Pennsylvania	$ 858.8	$ 821.2
Ohio	$ 774.8	$ 731.5
Maryland	$ 614.2	$ 539.9
Massachusetts	$ 582.1	$ 475.7
Florida	$ 521.2	$ 506.0
Michigan	$ 478.5	$ 427.1
Connecticut	$ 476.9	$ 434.4
Georgia	$ 445.1	$ 387.5
Arizona	$ 412.2	$ 381.6

Source: Memo from April Marchese, Director, Office of Natural Environment to FHWA Division Administrators on March 23, 2007 Available online at: http://www.fhwa.dot.gov/environment/cmaqpgs/msgobsrec1.htm.

Data Collection Procedures

After using the project selection criteria identified above to identify a reasonable set of projects from the CMAQ Database, a combination of methods was used to gather missing or incomplete information on the selected projects. For example, Internet searches were conducted to find project-specific information that might be available online, as well as to obtain the contact information for the State or MPO CMAQ representative. In several cases, information on CMAQ projects was available from Transportation Improvement Program (TIP) reports, congestion mitigation reports, and posted evaluation reports online.

If data were still missing and accurate contact information could be obtained, CMAQ representatives at the State DOT or MPO were contacted. Representatives were provided information on the purpose of the evaluation project and the specific CMAQ ID numbers of interest, and were asked for reports on the emissions reduction methodology originally used to calculate the emissions benefit, project cost information, and any available project evaluations. A list of the State and local project sponsors consulted for backup information for this report is listed in Appendix A.

In several cases, initial projects for which data and methodologies were gathered were eliminated from the study. This occurred due to limited or incomplete documentation of assumptions, which made it difficult to determine how precisely the emissions benefits were calculated; or due to assumptions that appeared unusually high or low and did not appear to be representative of typical projects within the project category. In a couple of cases, it was found that the project sponsor incorrectly calculated emissions effects based on the documentation provided (i.e., due to a mathematical error, or improper conversion). In these cases, the reported values were corrected.

Once missing or incomplete CMAQ reports, TIP reports, and/or emissions calculations were provided by the local representative to supplement data from the CMAQ database, the data were entered into individual project "templates." These one-page project profiles are designed to capture in one place all the critical facts, such as calculated travel impacts, emissions factors used, and the non-Federal and Federal costs related to the example. The individual profiles for each of the CMAQ-funded projects analyzed in this report can be found in Appendix C.

In selecting projects for inclusion in this study, emphasis was placed on profiling to the fullest extent possible, the costs, impacts on congestion and air quality, and other benefits for each project. While the selected projects are intended to be representative of typical CMAQ-funded projects, the emissions effects and congestion effects estimated for these projects are not statistically significant indicators of the effects of projects that have been funded through the CMAQ program.

3.   IMPACTS OF PROJECTS ON AIR QUALITY AND CONGESTION

This section describes the reported impacts of the selected CMAQ projects on transportation emissions and congestion levels. The data reported in this section are based on the materials reported by the sponsors of CMAQ-funded projects, or by the State DOT or MPO responsible for reporting to FHWA. These estimates of project effects reflect project-specific factors and local conditions, such as typical vehicle trips lengths and factors that affect vehicle emissions rates (such as temperatures, vehicle fleet mix, and vehicle speeds and operating conditions). They often utilize data from past local studies reflecting local factors (e.g., park-and-ride lot utilization rates, transit ridership levels on new services).

While these data are generally the best estimates of expected emissions benefits available to the project sponsors, the data have some limitations that that should be noted. Specifically, the reported effects are forecasts of expected effects, typically based on sketch planning analysis methods. In most cases, the effects have not been validated based on before-and-after studies or other post-project evaluations. For some types of projects, such as bicycle and pedestrian projects and transit service amenities, it is difficult to predict effects, given limited scientific studies, analysis tools, and established approaches for estimating travel and emissions impacts. As a result, there is a fairly high degree of uncertainty in some of the results. Another limitation is that in many cases, State DOTs or MPOs reported emissions benefits only for pollutants of concern in the local area, such as ozone-precursors. Consequently, effects on emissions of carbon monoxide and particulate matter were not reported for many projects, even in cases where the projects would be expected to reduce emissions of these pollutants.

Overall findings are summarized below, followed by a brief discussion of the project impacts organized in major project category and subcategory groupings. For each project type, a table summarizing quantitative findings is accompanied by a commentary on findings and trends.

General Observations

Direct and Indirect Effects on Congestion and Mobility

Some CMAQ projects are designed to reduce traffic congestion and to minimize delay experienced by drivers. Traffic flow improvement projects - such as traffic signalization improvements, incident management programs, and intersection improvements - reduce recurring and/or nonrecurring traffic delay on the transportation system. Project sponsors used a range of different techniques, from simulation modeling to simplified sketch planning, to estimate changes in travel delay or speeds.

Although many traffic flow improvement projects are small in scope (e.g., an individual intersection improvement), targeted investments can yield significant improvements in roadway level of service and intersection performance in specific locations. Consequently, these projects can have a large impact on the daily travel conditions experienced by individual drivers in the area where the project is implemented. Moreover, on highly-traveled corridors, even small changes in travel speeds can result in substantial travel time savings when multiplied over thousands of vehicles. For instance, an intersection improvement project in Louisiana estimated that travel conditions would improve from Level of Service (LOS) F, reflecting heavy congestion, to LOS C, and would yield a reduction of 1,459.2 vehicle-hours of delay per weekday.

Projects that reduce vehicle travel may also have impacts on congestion, but these effects are generally not quantified, and the primary travel benefit of these projects is generally enhanced mobility and multimodal choices. Projects such as bike and pedestrian facilities, shared ride programs, travel demand management (TDM) programs, and transit improvements may reduce vehicle miles traveled (VMT) by passenger vehicles, particularly during peak periods, and, therefore, may contribute to reduced traffic congestion. Freight/intermodal projects are often designed to shift goods movement from trucks to rail, and thereby reduce congestion associated with truck traffic on corresponding freight corridors.

These individual projects, however, often have limited impacts on travel demand in specific corridors or on a region-wide basis. For instance, several projects examined (including vanpool projects, park and ride lots, bicycle and pedestrian projects, and transit service improvements) were estimated to reduce less than 200 vehicle trips each day. Reductions of this level of trips may not have measurable effects on traffic congestion. Moreover, changes in travel speeds and delay depend on the volume of traffic by time of day, and impacts are non-linear (i.e., a reduction in 1,000 cars will not necessarily have twice the effect of a reduction in 500 cars on travel speeds). As a result, the magnitude of congestion relief due to VMT-reduction projects is difficult to predict or assess. There also are no standardized and simple methodologies to assess these effects. Given the lack of a specific funding for this type of analysis, and other demands placed on the program, it is perhaps not surprising that quantitative data on congestion benefits are very limited.

A primary purpose of bicycle and pedestrian projects, share ride programs, TDM programs, and transit service improvements is to enhance mobility by allowing greater travel choices. Over the long term and in combination other projects, projects such as bicycle paths and transit shuttles may improve mobility further by supporting transit-oriented development, an improved pedestrian environment, and enhanced multi-modal choices.

Some eligible CMAQ projects will not have any effects on traffic congestion or mobility. For instance, diesel engine retrofits and bus replacements are designed to reduce emissions rates from on-road vehicles without changing travel patterns. Similarly, dust mitigation projects are designed to reduce wind-blown dust on roadways without any changes in traffic congestion or mobility.

Direct and Indirect Effects on Air Quality

Overall, the analysis of selected projects suggests that emissions reductions have been achieved across the wide range of projects funded through the CMAQ Program, and that ultimately, each project helped contribute to some extent toward air quality goals. CMAQ projects can help reduce emissions through:

1) improving traffic flow, thereby reducing vehicle idling and stop-and-start driving conditions that are associated with higher levels of emissions;

2) encouraging changes in travel behavior that reduce motor vehicle miles traveled (such as shifts to ridesharing, transit, bicycling, or walking); and

3) using technologies to reduce the rate of emissions (such as through purchases of cleaner buses, or retrofits of diesel vehicles).

Given the small scale and localized nature of many CMAQ projects (e.g., a park-and-ride lot, a bicycle path, a transit shuttle), many CMAQ projects only yield small direct reductions in motor vehicle pollution. Among the projects reviewed in this study, the majority had emissions reduction estimates of less than 50 kg per day of both VOC and NOx, and less than 500 kg per day of CO. Although these estimated reductions are generally quite small, the combined effect of many small projects and those that are more regional in nature may help in achieving regional air quality goals. In fact, a number of regions take emissions reduction credit for regional demand management programs and other CMAQ-funded projects as part of their regional conformity analyses.

Moreover, the combined effect of many similar projects may help to achieve longer-term and more substantial indirect benefits to air quality. For instance, by contributing to development of a more multi-modal transportation system, by supporting access to transit, and by focusing attention to operational strategies, CMAQ projects can help support longer-term changes in travel behavior, land use, and attitudes toward transportation that support air quality goals and other related planning goals. These effects are very difficult to assess, and are not quantified for purposes of reporting to FHWA.

Additional Considerations on Air Quality Impacts

It should be noted that although project sponsors reported estimates of emissions benefits in the CMAQ database, and a consistent metric of kg per day is used, there are limitations associated with reporting a single emissions figure for each pollutant for each project. In considering the overall benefits of CMAQ projects on air quality, it is important to consider the following factors:

- **Days of Effectiveness** – Some projects have impacts every day of the year, while others only have effects on weekdays (such as projects that affect peak period traffic), and others have effects on even fewer days. For instance, bicycle projects might only be effective in encouraging shifts from driving during days when the weather is mild (for instance, the analysis of a bike path in Indiana assumed use 132 days per year), and analysis of a dust mitigation project that involved use of de-icing chemicals rather than sand would only be effective during winter months. The analysis of an ozone action days program in Rhode Island reported emissions effects based on changes in transit ridership due to a free transit program, and noted that the free transit days occurred on 4 days in 2005. Consequently, the reported emissions benefits in the CMAQ database are not sufficient to compare the impacts of projects at a national level. Many project sponsors estimate emissions benefits on an annual basis, in addition to daily effects, and this is particularly important to States and MPOs that rank projects on the basis of effects or cost-effectiveness as part of their selection process.

- **Duration of Benefits** – Some project benefits are expected to occur in the short-term, such as operational programs, like a ridesharing program or travel demand management incentives program. Other project benefits may have longer lasting impacts, notably infrastructure projects, like park-and-ride lots, transit rail, and bicycle and pedestrian facilities, which would be expected to last for perhaps more than 10 years and continue to generate emissions benefits over this time period.

- **Changes in Effectiveness over Time** – Since the CMAQ database only requires reporting of one emissions figure for each pollutant, emissions benefits were typically calculated or reported for one year in time. However, in reality, the stream of emissions benefits for a project is not likely to be constant over time. Overall, emissions rates from motor vehicles are declining, and so a project that produces a near constant travel impact, such as a park-and-ride lot or transit shuttle service (which are capacity constrained), is likely to have declining emissions benefits over time. On the other hand, some projects, such as regional employer trip reduction programs, might achieve increasing benefits over time as population and congestion in a region grow. Although not reported to the FHWA CMAQ database, some project sponsors estimated a stream of benefits over time, which is useful for purposes of project ranking and selection.

The following sections summarize congestion and emissions benefit findings for the projects reviewed in this study by project category.

Traffic Flow Improvements

Traffic flow improvements are designed specifically to meet the dual goals of the CMAQ program: decreasing congestion and reducing air pollution. In this report, traffic flow improvements are broken into three subcategories:

- Traffic Signalization and Intersection Improvements;

- Freeway Management; and

- High-Occupancy Vehicle Lanes.

Examining the emissions impacts of these strategies typically involves estimating travel speeds with and without the improvement in order to develop two different emissions factors for each situation. These emissions factors are then applied to VMT along the facility. In some cases, emissions benefits are calculated by estimating the reduction in vehicle delay and applying an idle emissions factor (grams per hour).[22] Some of these project analyses also account for changes in vehicle volumes associated with the improvements.

Traffic Signalization and Intersection Improvements

Seven CMAQ-funded traffic signalization and intersection improvement projects were reviewed in this analysis; quantitative cost and emissions findings are summarized in the table below.

STATE	CMAQ FUNDING	TOTAL COST	PROJECT TITLE	YEAR Funded	VOC (kg/day)	CO (kg/day)	NOx (kg/day)	PM10 (kg/day)	PM2.5 (kg/day)
Michigan	$660,000	$660,000	Signal Timing along Ryan Rd.	2002	-40.1	NR	NR	NR	NR
Louisiana	$4,400,000	$5,500,000	Continuous Flow Intersection at Airline and Sherwood Forest	2004	- 20.1	NR	- 5.2	NR	NR
Kentucky	$320,000	$400,000	Fiber Optic Cable Installation for Traffic Signal Optimization	2005	- 33.5	- 378.0	- 9.1	NR	NR
Ohio	$355,302	$639,543	Signal Timing along West Main Street	2005	- 5.1	-90.7	- 3.9	NR	NR
Tennessee	$33,000	$33,000	Signal Timing on SR-169 from Cedar Bluff to College St.	2005	- 15.0	NR	+ 2.2	NR	NR
Kentucky	$400,000	$500,000	Installation of Reversible Lanes on Nicholasville Road (US 27)	2006	- 2.9	-45.0	-1.1	NR	NR
New York	$2,000,000	$4,870,000	Construction of a two lane roundabout at Fuller St. and Washington St.	2007	-24.2	-24.2	-1.9	NR	NR

NR – Values were not reported by the local project sponsor or State DOT in the CMAQ database or other materials for the project.

---

[22] Calculated based on the emissions factor (in grams per mile) at 2.5 miles per hour, the slowest speed in MOBILE6, multiplied by 2.5 miles per hour, to generate an idle emissions factor in grams per hour.

Congestion/Mobility Benefits

Traffic signalization and intersection improvement projects are typically designed to reduce traffic congestion, increase travel speeds, and/or reduce delay. Congestion benefits were reported for each of the signalization and intersection improvement projects in the study, as described briefly below:

- Interconnection and modernization of 15 traffic signals along an urban minor arterial (Ryan Road) in Macomb County, Michigan, which borders the City of Detroit to the South and Lake St. Clair to the east, was estimated to improve travel speeds by 4 mph in both peak and off-peak periods, due to reduced delay at intersections.

- A modification to two intersections in East Baton Rouge Parish (Airline Highway – US 61 – and Siegen Lane/Sherwood Forest Boulevard) was undertaken to increase traffic flow and reduce congestion and delay. The intersections were operating at level of service F during peak hours, and a new design called a Continuous Flow Intersection would improve traffic operations at the intersection to acceptable levels of service. Simulations conducted using the VISSIM Microscopic Simulation Model, which estimates average delay in seconds per vehicle for each approach to the intersections, showed that the proposed improvements would enhance the throughput at the intersection for two hours during the morning peak period and two hours during the evening peak period while reducing delay time. For instance, in the evening peak at the intersection of Siegen/Sherwood and Airline, existing conditions were 6,200 vehicles per hour each experiencing an average of 178.3 seconds of delay, for a total of 368.5 vehicle hours of delay per peak hour; with the improvements, it was estimated that 6,700 vehicles per hour would experience an average of 34.4 seconds of delay, for a total of 64.0 vehicle-hours of delay per peak hour. In total, the two components of the intersection would reduce delay by 299.3 vehicle-hours in each morning peak hour and 429.9 vehicle hours in each evening peak hour, for a total of 1,459.2 vehicle hours saved per weekday.

- In Lexington, Kentucky, the installation of fiber optic cable for traffic signal optimization for the arterial highway network was estimated to reduce delay by 4 minutes per vehicle (determined by using an average reduction for a sample of 18 intersections), resulting in an estimated 6,360 vehicle hours of delay per day saved throughout the entire network.[23]

- Installation of coordinated signalized intersections to replace stop control at several intersections along Main Street in the City of Newark, Ohio was estimated to reduce delay by 702 vehicle hours per day at four of the main intersections involved in the project, based on analysis using a traffic simulation model.

- Tennessee DOT estimated an increase in average speed of 34 mph to 38 mph after traffic signal timing synchronization for a roadway in Knoxville, affecting 25,935 average daily vehicle trips.

- Installation of reversible lanes along Nicholsonville Road (US 27) in Fayette County, Kentucky, to allow three northbound lanes during the morning peak period was estimated to result in a 17 percent reduction in delay, or 63 vehicle hours saved during the morning and evening peak hour each day. According to the region's Congestion Management System report, for approximately 1.5 hours each morning, a queue of traffic bound for Lexington extended for a distance of nearly one-half mile, and often further due to incidents or inclement weather. The project would take advantage of unutilized median space and low early morning left turning volumes to create a

---

[23] Hours of delay was not reported directly by the project sponsors, but was calculated by the study team based on information provided in the project sponsor's emissions analysis.

third northbound traffic lane in the morning peak period by reassigning one of the left turn lanes on each side of an intersection as a through lane during this period.

- In Albany, New York, conversion of a signalized intersection into a roundabout was estimated to increase average speeds from 15 mph to 29 mph, and affect 48,670 vehicles over a quarter mile. These figures were calculated based on changes in vehicle delay, which were estimated to fall from an average of 31 to 47 seconds per vehicle at each approach to an average of 6 to 16 seconds.

Emissions Benefits

In general, traffic signalization and intersection projects that reduce vehicle delay will reduce emissions across all types of pollutants. However, traffic flow projects that increase travel speeds may have different effects on different pollutants. VOC emissions generally decline with increasing speeds, while CO and NOx emissions can begin to increase at speeds beyond 32 to 35 mph. As a result, some projects that increase speeds around certain ranges may actually increase CO and NOx emissions.

For the selected projects, the project sponsors estimated daily emissions reductions ranging from 2.9 kg to 40.1 kg of VOC. Daily NOx emissions reductions associated with each project show a smaller effect (from 1.1 kg to 9.1 kg reduced), with one project exhibiting a 2.2 kg increase in NOx emissions due to speed increases beyond 35 mph. CO reductions reported by sponsors indicate 24.2 kg to 378.0 kg emissions reductions each day.

None of the sponsors reported reductions in PM for these projects. Since EPA's MOBILE6 model does not account for the effects of changes in vehicle operating speeds on PM emissions, one would expect no reportable change in PM emissions for projects that alter vehicle operating speeds. Several project sponsors, however, calculated emissions benefits based on reduced vehicle idling time (e.g., calculating reduction in delay time due to the project and multiplying by idle emissions factors), in which case, PM emissions reductions could be calculated.

Costs

The total project costs for the signalization and intersection improvement projects ranged in magnitude from $33,000 to $5.5 million. The non-CMAQ share of project funding ranged from 0 to 20 percent of the total project cost. The total cost of signal timing projects will vary greatly depending on a number of local and project-specific factors, including the methods for coordinating signals, the number of signals included in the project, and the length of the roadway. The most expensive two projects both involved capital projects to redesign intersections. At $4.87 million and $5.5 million, respectively, the development of a continuous flow intersection in Louisiana and the construction of a roundabout in New York required substantially more funding than the signal timing projects. Although the capital costs are relatively expensive, the infrastructure and emissions benefits, associated with these types of projects could be long lasting.

Freeway Management

Four CMAQ-funded freeway management projects were documented in this analysis; quantitative findings are summarized in the table below.

STATE	CMAQ FUNDING	TOTAL COST	PROJECT TITLE	YEAR Funded	VOC (kg/day)	CO (kg/day)	NOx (kg/day)	PM10 (kg/day)	PM2.5 (kg/day)
Louisiana	$2,712,940	$2,712,940	ITS on I-10 from Acadian St. to Highland Blvd.	2003	-189.6	NR	-489.0	NR	NR
Washington	$998,037	$2,000,000	Duwamish ITS System	2004	-76.0	-939.0	-4.0	NR	NR
Connecticut	$1,279,246	$1,421,384	Incident Management System on I-95	2005	-6.1	NR	-3.00	NR	-0.004
Alabama	$240,000	$800,000	Alabama Service Patrols Program	2007	-31.3	NR	-11.9	NR	-0.12

NR – Values were not reported by the local project sponsor or State DOT in the CMAQ database or other materials for the project.

Freeway management projects improve traffic flow along major highways and often target travel impacts during peak periods when delays most often occur. These projects include service patrols that assist or remove disabled vehicles from blocking travel lanes, computer systems that control traffic flow onto freeways, monitoring devices that scan for incidents and provide motorist assistance or reroute traffic around the incident, and other Intelligent Transportation System (ITS) components. In most metropolitan areas, traffic incident-related delay (not including other non-recurring delay caused by weather, work zones, etc.) is estimated to account for between 25 and 30 percent of total congestion delay.[24] When vehicles are cleared from the motorway and/or other vehicles are alerted to incidents ahead, idling is reduced and traffic speeds along freeways can return to more optimal levels.

Congestion/Mobility Benefits

Freeway management projects can reduce recurring and/or nonrecurring delay associated with incidents. For example, an ITS system project in Seattle, Washington included interconnection of traffic signals and controller equipment upgrading, installation of variable message signs and other driver information systems, and implementation of traffic control strategies to monitor traffic conditions and accidents. This project was designed to minimize the conflicts among freight movement, transit travel, commuter traffic, and ferry access, while enhancing safety and mobility for people and goods. The project sponsors estimated a 10 percent increase, or 2 mph, in both peak and off peak speeds due to the program.

In Connecticut, development of a 13.94 mile portion of an incident management system on I-95 included the installation of a fiber-optic communication system, video surveillance, traffic flow monitors, and a link to the Bridgeport Operations Center. The incident management project was designed to allow operational problems to be identified sooner and enable faster dispatch of the proper response equipment and medical services to a site. Based on data from the "Connecticut Freeway Management System" report, which reported effects for a 65 mile length corridor, it is estimated that this type of system will result in annual delay savings of 368,000 vehicle hours, assuming proportional benefits for the 13.94 mile corridor (based on an assumption of a congested incident speed of 5 mph, and a free flow speed of 55 mph).

---

[24] FHWA. 2003. Freeway Management and Operations Handbook.
http://ops.fhwa.dot.gov/freewaymgmt/publications/frwy_mgmt_handbook/toc.htm.

The Alabama Service and Assistance Patrol (ASAP) Program, an incident management program of the Alabama DOT and Alabama State Troopers, offers services to disabled motorists to reduce response time and to minimize major disruption of interstate traffic flow. This program was estimated to result in a savings of 3,849 vehicle hours of delay per incident for an estimated 111 incidents per year, resulting in a savings of over 427,000 vehicle hours per year. An advanced traffic management center in the Baton Rouge metropolitan area, including incident detection and response, motorist assistance, and surveillance components along I-10, also was designed to reduce incident-based delay, but did not report hours of delay reduced.

Emissions Benefits

As with other traffic flow improvements, freeway management projects that cause an increase in travel speeds may have varying effects on different pollutants, depending on the magnitude of the overall speed change. Emissions reductions reported by project sponsors for the four projects indicated daily VOC emissions reductions ranging from 6.1 kg to 189.6 kg per day, and daily NOx emissions reductions ranging from 4.0 kg to 489.0 kg per day. One project sponsor estimated a 939.0 kg CO emissions reduction; the three other projects did not report CO reductions. Two projects reported $PM_{2.5}$ emissions benefits of 0.004 and 0.12 kg per day. Due to reduced vehicle idling, one would expect reductions of CO and PM for each project.

It should be noted that the Louisiana project, which reported the highest emissions reductions, assumed that the incident detection and response, motorist assistance, and surveillance components of the ITS project along I-10 would result in a 4.41 percent reduction in total emissions for traffic along the I-10. This assumption appears to be somewhat high in comparison to the other analyses, and is based on data showing that 4.9 percent of freeway emissions are associated with nonrecurring congestion, and an assumed 90 percent reduction in incident-based emissions. The 90 percent effectiveness factor is based on an effectiveness rate of 50 percent for incident detection and response, 25 percent for motorist assistance, and 15 percent for surveillance.

Costs

The total project costs of the selected projects ranged in magnitude from $800,000 to $2,712,940. In general, most freeway management projects involve major corridors or a network of roadways and so have substantial capital and/or operating costs. The non-CMAQ share of project funding ranged from 0 percent to about 50 percent of the total project cost.

High-Occupancy Vehicle (HOV) Lanes

Although the CMAQ program has helped to fund a number of HOV projects throughout the history of the program, only a small number of HOV projects have used CMAQ funding since FY 2000, which is the focus of this representative analysis. Consequently, only one project was identified for analysis in this study – construction of an HOV interchange in Dallas. Quantitative findings are summarized in the table below.

STATE	CMAQ FUNDING	TOTAL COST	PROJECT TITLE	YEAR FUNDED	VOC (kg/day)	CO (kg/day)	NOx (kg/day)	PM10 (kg/day)	PM2.5 (kg/day)
Texas	$17,152,000	$254,570,093	Dallas HOV Interchange	2002	-68.8	NR	-135.3	NR	NR

NR – Values were not reported by the local project sponsor or State DOT in the CMAQ database or other materials for the project.

An HOV lane is a travel lane usually reserved for use by vehicles with more than one occupant, such as carpools, vanpools and buses, during peak periods or longer periods. They are often located next to the regular or general purpose lanes. Most users of HOV facilities can expect a substantial savings in travel time, as well as a commute time that is more reliable and predictable on a daily basis. Because HOV lanes carry vehicles with a higher number of occupants, the amount of vehicles needed to transport those occupants is reduced, resulting in fewer vehicle trips and lower overall VMT.

There are approximately 126 HOV freeway projects in 27 metropolitan areas in the U.S. These HOV facilities include over 1,000 miles of roadway, most often on interstate freeways. HOV lanes have also been implemented on arterial roads, especially those related to bus-only applications.[25]

Congestion/Mobility Benefits

HOV lanes improve mobility for people who choose to rideshare and for transit users by allowing a faster trip compared to being in general purpose lanes. HOV lanes also offer congestion benefits primarily by encouraging more passengers to travel in fewer vehicles, and can provide more person throughput on a fixed amount of transportation infrastructure. Additionally, some States open HOV lanes to unrestricted traffic during non-peak hours, increasing the overall capacity for vehicle movement.

In Dallas, the HOV interchange project sponsors estimated that 2,929 daily vehicle trips would be reduced due to the HOV facility, based on an estimate that 56 percent of transit and rideshare users on the facility previously drove alone. Effects on overall congestion levels and speeds on the highway were not quantified, although the calculation of emissions effects took into account the differences in speeds between the general purpose lanes and HOV lane.

Emissions Benefits

HOV lanes affect air pollution emissions in several ways. First, restricting the lanes to certain vehicles encourages ridesharing among commuters and results in fewer vehicle trips and an overall reduction in emissions of all pollutants. HOV lanes also increase travel speeds for HOV traffic, and sometimes along the entire roadway. Increases in travel speeds, as noted previously, will have different effects for different pollutants depending on the magnitude of the increase.

The Dallas HOV interchange project was estimated to reduce 68.8 kg of VOC and 135.3 kg of NOx per day. CO and PM reductions were not reported for this project, but might occur due to the reduction in vehicle travel. The calculation accounted for both a reduction in VMT due to people shifting to transit and

---

[25] FHWA. "Frequently Asked HOV Questions" http://ops.fhwa.dot.gov/freewaymgmt/faq.htm#faq7

ridesharing, and an increase in vehicle speeds for the traffic shifting from general purpose lanes to the HOV lane. The analysis did not take into account the potential speed changes that occur for the vehicles remaining in the general purpose lanes.

Costs

The construction of a new HOV lane and the ramps and other infrastructure required for an HOV system can be expensive. The total public cost of the reviewed project was $254,570,093, but the CMAQ program only paid for approximately 7 percent of the total project cost. The bulk of funding came from National Highway System (NHS) funding, along with a lesser amount from State and local sources. Although HOV projects incur large capital costs upfront, the infrastructure and corresponding emissions benefits, may be long-lasting.

Shared Ride Programs

Shared ride programs encompass a wide variety of projects that focus on changing travel behavior to reduce air pollutant emissions from light-duty vehicles. These programs offer services that encourage single-occupant vehicle travelers to group rides with other travelers, generally in carpools or vanpools, thus increasing the average number of occupants per vehicle trip and reducing total vehicle trips and VMT. Projects analyzed include:

- Regional Ridesharing Programs;

- Vanpool Programs; and

- Construction of Park and Ride Lots.

Regional Ridesharing

Three CMAQ-funded regional ridesharing programs were documented in this analysis.

STATE	CMAQ FUNDING	TOTAL COST	PROJECT TITLE	YEAR FUNDED	VOC (kg/day)	CO (kg/day)	NOx (kg/day)	PM10 (kg/day)	PM2.5 (kg/day)
Maryland	$956,000	$956,000	11 County Ridesharing Program Operations	2002	-35.0	NR	-110.0	NR	NR
Pennsylvania	$480,000	$600,000	University of Pittsburgh TDM Program	2005	-26.2	-187.4	-30.9	NR	NR
Alabama	$700,000	$700,000	CommuteSmart Commuter Services Program Operations	2007	-10.2	NR	-12.0	NR	-0.1

NR – Values were not reported by the local project sponsor or State DOT in the CMAQ database or other materials for the project.

Regional ridesharing programs provide ride-matching services, employer outreach, and incentives to commute by carpool or vanpool. These incentives can include free gas cards, award programs, and travel subsidies. Ride-matching may encourage people to establish regular carpool routines or can be dynamic and create systems to match individuals who want to travel to/from similar locations in real-time. These programs largely serve a supportive or facilitative role, and help to optimize use of existing transportation infrastructure and services. Their success depends, in part, on the commute options existing in the community, such as HOV lanes and transit services.

Congestion/Mobility Benefits

Regional ridesharing programs can improve mobility by giving people greater options in meeting their travel needs, and can reduce travel costs for people who choose to rideshare. The congestion benefits of a regional ridesharing program will depend on the number of new carpools and vanpools that are formed and the extent to which participants previously drove alone. It should be noted, however, that if some of the persons who choose to rideshare previously rode transit, this mode switch would not necessarily be beneficial to congestion or emissions. Reductions in VMT are also dependant on the length of the trips, and the length of the carpool trip to pick up riders.

The Birmingham MPO estimated that its regional CommuteSmart Commuter Services Program – which includes a ridesharing database, a vanpool program with 34 vans in 2007, and a carpool program – would result in a reduction of about 312 vehicle trips per weekday. The primary users of these services have longer-than average commutes. At an average one-way trip length of 39.5 miles, the program reduces a total of 9,470 vehicle miles of travel per weekday.

The TDM program in the Oakland area of Pittsburgh, Pennsylvania surrounding the University of Pittsburgh, offers ridesharing coordination, employer-sponsored vanpools, and carpool programs. This program was estimated to reduce 2,024 vehicle trips per day at an average one-way trip distance of 5.45 mile, for a total of 22,062 vehicle miles reduced per day.

The Maryland program - which funds a ridesharing program in eleven counties in the Baltimore and Washington, DC metropolitan areas - was estimated to reduce about 3,000 vehicle trips per weekday, based on data showing 12,360 rideshare applicants in the programs and an estimate that 24 percent will take part in ridesharing each day. At an average one-way trip distance of 14 miles, this program results in about 84,000 vehicle miles reduced per day.

None of the project sponsors for the three projects submitted information on delay reductions or travel speed improvements anticipated with the projects.

Emissions Benefits

By encouraging people who would normally drive alone to share trips, ridesharing programs reduce motor vehicle travel and associated emissions. Daily emissions reductions associated with the selected projects range from 10.2 kg to 35.0 kg of VOC and from 12.0 kg to 110.0 kg of NOx. One project was estimated to result in a CO emissions reduction of 187.4 kg per day. $PM_{2.5}$ emissions effects were reported in the analysis of one project, showing a reduction of 0.1 kg per day. However, since these projects reduce VMT, all three projects would likely reduce emissions of all pollutants.

Costs

The CMAQ program is a key funding source for many regional ridesharing programs, with grants used to cover operating expenses, such as advertisements, outreach materials, and commute incentive purchases. The total public cost of these projects ranged in magnitude from $600,000, for two years of the Pittsburgh program ($300,000 per year) to $956,000 for the annual costs of the eleven-county Maryland program. The non-CMAQ share of project funding ranged from 0 to 20 percent of the total project cost.

VANPOOL PROGRAMS

Three CMAQ-funded vanpool programs were reviewed in this analysis. Findings are summarized in the table below.

STATE	CMAQ FUNDING	TOTAL COST	PROJECT TITLE	YEAR FUNDED	VOC (kg/day)	CO (kg/day)	NOx (kg/day)	PM10 (kg/day)	PM2.5 (kg/day)
Utah	$448,000	$448,000	15 new vans for vanpool leasing program	2002	-12.2	-136.9	-14.9	NR	NR
Utah	$148,866	$180,866	5 new vans for vanpool leasing program	2005	-3.2	-37.2	-4.0	NR	NR
Kentucky	$96,000	$120,000	6 new vans for LexTran Vanpool	2006	-10.4	-80.2	-5.3	NR	NR

NR – Values were not reported by the local project sponsor or State DOT in the CMAQ database or other materials for the project.

Vanpool programs provide vehicles that are owned by an organization or public agency to commuters who live in a common geographic area and who share an employment destination. Each vanpool carries between seven and fifteen passengers on a van or bus, operates on weekdays, and typically travels between one or two common pick-up locations and the place of work. The vehicles may be operated by a paid driver or by the commuters themselves, depending on local or program preferences. Employers or institutions frequently enable vanpool operations in any of a variety of supportive or financial ways.

Each of the selected projects involved the purchase of passenger vans: 15 new 8-passenger vans for the Utah Transit Authority (UTA) Vanpool Leading Program in 2002 to be used in the Salt Lake City and Ogden areas; five new 8-passenger vans for the UTA Vanpool Leasing Program in 2005 to be used in the Ogden and Layton area; and six new 12-passenger vans for LexVan, a commuter vanpool program managed by the Lexington Bluegrass Mobility Office in Kentucky.

Congestion/Mobility Benefits

Vanpool projects can improve mobility by giving people an option to meet their commuting needs at lower cost than driving alone. The congestion benefits of vanpool programs will vary depending on the number of vanpools established through the program, the number of passengers, and the length of a trip. Typically, vanpools are successful in areas with longer commutes and where they utilize established park and ride lots as the common pick up location. For small vanpool programs serving a limited number of passengers, the net reduction in vehicle trips is small: the three reviewed projects were estimated to remove from 40 to 120 drivers from the road each day, and reduce overall VMT by 3,000 to 6,600 vehicle miles traveled per day. Consequently, congestion benefits would be too difficult to quantify. These projects, however, can result in important benefits to individual passengers, including reduced fuel and vehicle maintenance costs, and improved quality of life due to reduced commuting stress and time that can be spent reading or in other activities during the vanpool trip.

Emissions Benefits

Vanpools reduce VMT on the roads, and therefore should reduce emission of all pollutants. Although the vanpool vehicle may produce more emissions than an individual automobile, the emissions are considerably less than the total of the seven to fifteen individual vehicle trips that are typically replaced. Among the three projects, estimated daily VOC emissions reductions associated with each project ranged from 3.2 kg to 12.2 kg, daily NOx emissions reductions ranged from 4.0 kg to 14.9 kg, and daily CO reductions ranged from 37.2 kg to 136.9 kg.

Costs

CMAQ funding for vanpool programs may be used for capital costs, such as purchase of new or replacement vans, or for operating expenses, such as paid advertisements and printing outreach materials. For these three projects, CMAQ funding was used for the purchase of additional vans and does not include any operating costs. The total public cost of the selected projects ranged in magnitude from $120,000 to $448,000. The non-CMAQ share of funding ranged from 0 percent to 20 percent of the total project cost, and in some cases, included the vanpool fares.

Park and Ride Lots

Five CMAQ-funded park and ride lot projects were reviewed in this analysis.

STATE	CMAQ FUNDING	TOTAL COST	PROJECT TITLE	Year Funded	VOC (kg/day)	CO (kg/day)	NOx (kg/day)	PM10 (kg/day)	PM2.5 (kg/day)
Maryland	$132,817	$132,817	Two new 25 space lots	2000	- 0.01	NR	-0.06	NR	NR
Wisconsin	$48,000	$48,000	Lake Geneva and Root Creek Lots	2000	- 1.5	NR	- 3.8	NR	NR
Maryland	$1,218,831	$1,218,831	MD 210 and MD 373 500 space lot	2002	- 1.4	NR	- 5.9	NR	NR
Kentucky	$844,800	$1,056,000	Walton/Union Lot with 200 spaces	2005	-0.9	- 33.8	- 3.2	NR	-0.1
Washington	$4,150,000	$20,000,000	Expansion of Terrace Station Transfer Lot to 880 spaces	2005	- 18.0	- 145.0	- 9.0	NR	NR

NR – Values were not reported by the local project sponsor or State DOT in the CMAQ database or other materials for the project.

Park and ride lots are transportation facilities that provide people a secure location to park their vehicles before joining a carpool, vanpool, or transit service. Typically located in suburban areas, these projects provide commuters the flexibility of driving to a central location near their home and then completing the majority of their commute using transit or ridesharing. The selected projects range widely in size and include:

- Construction of two new 25-space lots in Cecil County, Maryland;

- Construction of 300 spaces at two lots in Southeastern Wisconsin;

- Addition of 500 spaces at an existing lot in suburban Maryland;

- Development of a new 200-space lot, along with improvements to existing lots, including improving signage, adding bicycle parking racks, and providing information kiosks in Kentucky; and

- Construction of a new multi-level parking structure over an existing park-and-ride lot in Seattle, Washington, increasing capacity from 388 to 880 spaces.

Congestion/Mobility Benefits

The congestion benefits of park and ride lots are associated with reductions in freeway and arterial VMT during peak periods when commuters use the park and ride lots. The reductions are dependent on the number of spaces that will be created as part of the project, and the utilization of the available spaces.

The projects generally reported 126 to 738 vehicle trips reduced per day (the exception is the small park and ride project in Cecil County, Maryland). More precisely, vehicle trips are not eliminated since users still drive to the lot; these trips are shorter in length since commuters drive to the park and ride facilities. Since carpool trips, however, tend to be longer than average regional vehicle trip lengths, VMT reduction typically is larger than for other types of programs affecting a similar number of trips (e.g., bicycle projects, bus services).

Emissions Benefits

These projects improve air quality by reducing the number of vehicle miles traveled each day. Because motorists are required to drive to the park and ride lots, these projects will not reduce the number of vehicle cold starts, when the highest levels of CO, NOx, and VOCs are emitted.

Estimates of daily VOC emissions reductions associated with each project ranged from 0.01 kg to 18.0 kg. Daily NOx emissions reductions associated with each project ranged from 0.06 kg to 9.0 kg.

It should be noted that the project with the smallest impacts (two park and ride lots in Maryland funded in 2000) only involved the addition of 50 parking spaces, and assumed a very low utilization rate (15 percent) and a low percentage of users who are new riders (15 percent). The 2002 Maryland park and ride project used more typical assumptions, estimating that 56 percent of spaces would be utilized and 45 percent would be new riders; using similar assumptions for the 2000 park and ride project would result in emissions estimates approximately 11 times larger (e.g., -0.13 kg/day VOC, -0.65 kg/day NOx). Park and ride lot projects would be expected to reduce emissions of all motor vehicle-related pollutants. CO reductions reported by two projects indicate 33.8 kg to 145.0 kg emissions reductions each day, and $PM_{2.5}$ emissions reductions were reported by one project in Kentucky to be 0.1 kg each day.

Costs

CMAQ funding is usually provided as a portion of the total cost of construction of new facilities or the expansion and/or resurfacing of park and ride lots. The total public cost of these projects ranged widely, from $48,000 for two park and ride lots in Wisconsin to $20 million for construction of a multi-level parking garage in the Seattle region. In the case of the $20 million project, $4.15 million in capital costs were funded through CMAQ over two different funding years.

Travel Demand Management

Travel demand management (TDM) programs typically focus on reducing the number of vehicle trips by commuters during peak hours. TDM strategies are often linked to employer-based strategies and include encouragement of alternative work schedules, telework programs, guaranteed ride home initiatives, and Ozone Alert Days. They also may involve regional marketing efforts to support transit, ridesharing, and other travel options.

Four CMAQ-funded TDM projects were reviewed in this analysis, two of which are part of the Metropolitan Washington Council of Governments' (MWCOG) Commuter Connections Program.

STATE	CMAQ FUNDING	TOTAL COST	PROJECT TITLE	Year Funded	VOC (kg/day)	CO (kg/day)	NOx (kg/day)	PM10 (kg/day)	PM2.5 (kg/day)
Colorado	$73,000	$91,250	Coordinate Telework Program	2001	-2.0	-14.0	-2.0	NR	NR
DC, Maryland, Virginia	$9,000	$15,000	Regional Employer Outreach, Bicycles	2002	-1.0	NR	-1.0	NR	NR
DC, Maryland, Virginia	$772,110	$1,678,500	Regional Guaranteed Ride Home Program	2005	-95.2	NR	-216.8	NR	NR
Rhode Island	$168,000	$168,000	Ozone Alert Days	2005	-23.0	-251.3	-26.5	NR	NR

NR – Values were not reported by the local project sponsor or State DOT in the CMAQ database or other materials for the project.

TDM strategies often are implemented through directing marketing, services, and informational tools to encourage the use of available travel options. Commuters frequently are the focus of TDM actions because of their regular, predictable driving patterns, the possibilities of employer partnerships, and expanded opportunities for ridesharing programs.

Congestion/Mobility Benefits

Travel demand management programs improve mobility by supporting a range of travel options, including not only choices of alternatives modes to driving alone, but also telecommuting and changes in work schedules to avoid travel during peak period hours. The congestion benefits of TDM strategies can be attributed to shortened vehicle trips, the shifting of peak-period trips to non-peak hours, and the elimination of trips altogether. For the four selected projects, estimated vehicle trip reductions ranged from 125 vehicle trips per day for the Washington, DC-area bicycle outreach effort to 12,350 vehicle trips per day for the region's Guaranteed Ride Home Program. Since these programs are part of an integrated TDM program involving multiple elements, and credit is being taken for this program as a Transportation Emissions Reduction Measure (TERM) as part of the region's conformity determination, the regional MPO (MWCOG) has taken care to analyze the impacts utilizing surveys and other tracking data. As with other types of VMT reduction programs, impacts on travel speeds are generally difficult to assess and were not quantified by the project sponsors.

Emissions Benefits

Emissions reductions estimated by project sponsors were generally small for the selected projects, with the exception of the Washington, DC region's Guaranteed Ride Home program. Daily VOC and NOx emissions reductions associated with two of the projects were at or under 2.0 kg/day; in the case of the DC region's Guaranteed Ride Home program, emissions reductions were estimated at 95.2 kg/day of

VOC and 216.8 kg/day of NOx reduced. The Rhode Island program is an example of an "episodic" type program which is not in effect every day, but only on occasions when an ozone alert day is called. Hence, its benefits – associated with a fare free transit program on ozone alert days – only accrue on those few days a year when these events occur.

Costs

CMAQ funding is usually provided as an operating subsidy for TDM strategies. The total public cost of these projects ranged from $15,000 for the Washington, DC regional outreach on bicycling to more than $1.67 million programmed for the regional Guaranteed Ride Home program, including the costs of marketing, payment for rides, and staff labor. Total funding for the Commuter Connections program has ranged from $4.28 million to $5.11 million annually over the period FY 2002 to FY 2008, and includes seven related TDM program elements: Metropolitan Washington Telework Resource Center (TRC), Expanded Telecommuting, Guaranteed Ride Home, Integrated Rideshare, Employer Outreach, Employer Outreach for Bicycling, and Mass Marketing (a large-scale, comprehensive media campaign). Funding comes from Maryland, Virginia, and the District of Columbia, of which CMAQ funding makes up at least half.

Bicycle/Pedestrian Facilities

Bicycle and pedestrian projects and programs include a wide range of investments and strategies to facilitate and encourage non-motorized travel. Some examples of these projects include bicycle paths and lanes, sidewalks, bicycle racks or lockers, pedestrian urban design enhancements, bicycle/pedestrian marketing materials, and bicycle sharing projects.

Four CMAQ-funded bicycle and pedestrian projects were reviewed in this analysis.

STATE	CMAQ FUNDING	TOTAL COST	PROJECT TITLE	Year Funded	VOC (kg/day)	CO (kg/day)	NOx (kg/day)	PM10 (kg/day)	PM2.5 (kg/day)
Massachusetts	$639,008	$1,300,000	8.3 mile Swansea Bikeway Facility	2002	-0.5	-3.0	-1.1	NR	NR
Indiana	$1,600,000	$2,000,000	4.3 mile Bike Path to Pinhook Park	2005	-0.4	-2.7	-0.5	NR	NR
Colorado	$63,910	$600,000	Construction of a Transit Bike Depot	2006	-0.9	-6.7	-0.9	NR	NR
New York	$2,400,000	$3,000,000	NYC CyclistNET Marketing Program	2007	-2.4	-38.4	-2.0	-0.9	-0.04

NR – Values were not reported by the local project sponsor or State DOT in the CMAQ database or other materials for the project.

Bicycle and pedestrian projects often serve multiple goals, including improving mobility and safety. By providing bicycle and pedestrian access across barriers such as arterial roads, freeways, and/or train tracks, these projects can not only substitute for driving trips but also can improve mobility and access for non-drivers. Projects can also improve the safety of walkers and bicyclers by filling in gaps on existing, planned, or proposed routes and addressing potential hazards in existing facilities. Non-motorized forms of transportation require no fossil fuels, and are often considered in the context of goals such as sustainability, reducing energy consumption, and reducing greenhouse gas emissions.

Congestion/Mobility Benefits

Bicycle and pedestrian projects can contribute to improvements in mobility by providing additional options for people who might choose walking or biking. These projects improve the ability to reach desired goods, services, activities and destinations using non-motorized forms of transportation and may help diminish the need for automobile travel.

Bicycle and pedestrian projects may reduce congestion to the extent they shift mode choice from single occupancy vehicles to bikes and walking, and often are more successful in reducing VMT in locations where short driving trips, such as trips to local shopping areas, schools, or commercial districts, are common. While bicycle and pedestrian projects can reduce vehicle trips during both peak and off-peak times, congestion benefits are usually limited due to the relatively short distances of trips and to seasonal limitations on bicycling in some areas. The four projects reviewed had estimated reductions of 83 to 902 vehicle trips per day, with the largest figure reported for the New York bicycling program. Given the relatively small impacts at reducing vehicle travel, the bicycle and pedestrian projects assessed for this study did not provide estimates for changes in speed or delay times on the system.

Emissions Benefits

Bicycle and pedestrian projects generally have modest effects on emissions. Typically, pedestrian trips have a maximum distance of 1 mile and bicycle trips a limit of 5 miles, which reduces the ability of these projects to substitutes for driving for many commuters. Bicycle and pedestrian projects may be more effective when designed to enhance access to transit, so that longer trip lengths may be reduced.

Project sponsors generally estimated small reductions in motor vehicle emissions – typically under 1.0 kg/day for VOC and NOx. CO reductions reported by sponsors indicate a range of benefits from 2.7 kg to 38.4 kg emissions reductions each day. PM_{10} emissions reductions were reported by one project to be 0.9 kg each day. The same project reported daily $PM_{2.5}$ emissions reductions of 0.04 kg. All of the projects, however, would be expected to reduce PM_{10} and $PM_{2.5}$ from motor vehicle exhaust.

Costs

CMAQ funding is usually provided for capital improvements, but can also be an operating subsidy for the operation of marketing or bike sharing programs. The total public cost of these projects range in magnitude from $600,000 to $3 million. The non-CMAQ share of project funding ranged from 0 percent to 89 percent of the total project cost in the case of the Colorado bike depot (CMAQ costs reported were for architectural design and engineering documents to create the site design).

Transit Service Improvements

CMAQ funds may be used to support projects that increase the use of public transportation systems. Generally, there are three broad categories of transit service-related projects or programs: provision of new or expanded bus services, provision of new or expanded rail services, and service upgrades and rider amenities on existing transit services. Routine maintenance and rehabilitation of existing transit facilities are not eligible for CMAQ funding. However, substantial changes to transit stations or facilities that are likely to increase ridership and reduce emissions are eligible.[26]

New Bus Services

These strategies include the establishment of new routes, increased frequency of vehicles, expanded hours of operation, or increased coverage of routes. Three CMAQ-funded projects that provide new bus services were analyzed.

STATE	CMAQ FUNDING	TOTAL COST	PROJECT TITLE	Year Funded	VOC (kg/day)	CO (kg/day)	NOx (kg/day)	PM10 (kg/day)	PM2.5 (kg/day)
Wisconsin	$157,382	$196,727	City of Racine New Sunday Bus Service	2001	-2.9	NR	-3.2	NR	NR
New York	$264,000	$420,000	Expanded S92 Bus Route	2005	-6.7	-153.4	+7.2	+1.0	+1.0
Rhode Island	$440,000	$550,000	Expanded Route 30 and New Route 12	2005	-6.7	-191.0	-11.1	NR	NR

NR – Values were not reported by the local project sponsor or State DOT in the CMAQ database or other materials for the project.

Bus service improvement projects improve both air quality and congestion levels in the local community by increasing the use of transit services and reducing the number of auto trips. New bus routes make transit a more convenient transportation option and may reach areas of the community that were previously underserved or not served at all. Reductions in wait times for transit vehicles may lead to a faster overall trip for passengers, further increasing the number of transit users. Finally, increasing the hours of transit service along certain routes allows people to use the transit system at hours that were not previously available, thus allowing them more latitude in scheduling their trips and allowing for unforeseen changes in itinerary.

Congestion/Mobility Benefits

New bus service can reduce congestion by reducing vehicle trips and VMT. The extent of benefits will depend on several factors, including the extent to which new transit users drive to bus stops, the length of the new service, and the number of additional buses in operation in mixed-traffic. New bus services provide mobility improvements, to the extent that the services provide additional transportation options for users to choose. Mobility benefits will likely be greatest when land-use patterns and other supporting strategies, such as bicycle/pedestrian connections and rider amenities, are already in place. The projects selected for this study reduced between 72 to 358 vehicle trips per day, and project sponsors did not assess impacts on delay and travel speeds.

Emissions Benefits

Bus service improvements can reduce emissions of all pollutants by reducing the number of trips by single-occupancy vehicles and VMT. However, the new bus services also produce emissions, which may

---

[26] FHWA "Guidance on the Congestion Mitigation and Air Quality Improvement (CMAQ) Program Under the Safe, Accountable, Flexible, Efficient Transportation Equity Act: A Legacy for Users." Page 11.

offset some of the emissions reductions from personal vehicles and in some cases, NOx and PM emissions could increase due to emissions from the diesel engines in buses if the new services do not attract a sufficient number of new riders who previously drove. Emissions reductions reported by project sponsors indicate a range of anticipated benefits from implementation of these projects. Daily VOC emissions reductions associated with each project ranged from 2.9 kg to 6.7 kg. Daily NOx emissions reductions associated with each project ranged from 3.2 kg to 11.1 kg. Due to increased emissions from the new bus services, the New York project estimated a NOx emissions increase of 7.2 kg per day and PM_{10} and $PM_{2.5}$ increase of 1.0 kg/day. The other two projects did not account for the increase in emissions due to the new bus services, only the reduced emissions from personal vehicles.

Costs

CMAQ funding is usually provided for the operating costs associated with new bus services, but can also be available as a portion of the capital costs to purchase new buses. Only one of the projects, the S92 bus route on Long Island, included estimates of transit fares in determining project costs. The project sponsors estimated a total project cost of $420,000 and farebox revenues equal to $90,000, resulting in a net public cost of $330,000. The total public costs of the selected projects, without consideration for farebox revenues, ranged in magnitude from $196,727 to $550,000.

New Rail Services

New passenger rail services include establishing new routes, increasing the frequency of service, expanding the hours of operation, or the overall coverage of transit corridors. Three CMAQ-funded rail service projects were reviewed in this analysis.

STATE	CMAQ FUNDING	TOTAL COST	PROJECT TITLE	Year Funded	VOC (kg/day)	CO (kg/day)	NOx (kg/day)	PM10 (kg/day)	PM2.5 (kg/day)
Utah	$4,000,000	$4,000,000	Purchase of 5 New Light Rail Vehicles	2002	-27.0	-305.0	-33.0	NR	NR
Texas	$36,253,821	$70,472,342	TRE Double Tracking of Segments	2003	-67.2	NR	-110.0	NR	NR
Connecticut	$2,400,000	$3,000,000	Construct Rail Station Platforms and Bridge	2005	-6.0	NR	-6.0	NR	-1.0

NR – Values were not reported by the local project sponsor or State DOT in the CMAQ database or other materials for the project.

Projects to expand rail services can improve both air quality and congestion levels by reducing the number of auto trips, as well as bus transit trips, which may contribute to congestion and emissions. The projects include purchase of new light rail vehicles for the TRAX North/South line in Salt Lake City to enable additional services; double tracking segments of the Trinity Railway Express (TRE) commuter rail line between Dallas and Fort Worth to enable expanded capacity; and construction of a new commuter rail station along the Metro-North commuter rail line to serve Fairfield, Connecticut, including students of Fairfield University and nearby areas within the city of Bridgeport.

Congestion/Mobility Benefits

New and expanded rail services may provide mobility improvements in the form of increased transportation mode options for users in the community, and often will provide faster travel times than existing bus services. Improvements in mobility will likely be greatest when land-use patterns, inter-modal connections, and other supporting strategies, such as bicycle/pedestrian connections and rider amenities, are already in place. New or expanded rail services also may reduce congestion by attracting riders who previously drove their own vehicles. The congestion benefits will depend on several factors, including the extent to which new transit riders drive to the station, the length of vehicle trips reduced, and the existence of supporting land use patterns and bicycle, pedestrian, and parking access to stations. The three selected projects were estimated to reduce from 400 (Connecticut) to 5,400 (Dallas) vehicle trips per day. The project sponsors did not assess impacts on delay and travel speeds.

Emissions Benefits

New rail services and routes may reduce emissions of all pollutants by reducing VMT. These types of projects are often most effective when implemented in areas that have a large, established transit network. Daily VOC emissions reductions estimated by project sponsors ranged from 6.0 kg to 67.2 kg. Daily NOx emissions reductions ranged from 6.0 kg to 110.0 kg. CO emissions reductions were reported by one project sponsor as 305.0 kg per day. $PM_{2.5}$ emissions reductions were reported by one project sponsor to be 1.0 kg per day. These emissions effects only take into account the reduction in personal vehicle travel.

The emissions benefits of projects to provide new diesel rail services should include consideration of the increase in off-road emissions from operating locomotives. In the case of the Utah and Connecticut reviewed projects, there were no new diesel emissions, since these involved light rail and construction of a new rail station but no new service. The documentation for the Dallas project noted that there will not be any new emissions of NOx and VOC from diesel locomotives due to the double-tracking project;

however, presumably the calculation of emissions benefits accounts for new ridership associated with higher service levels.

Costs

All three of the projects had costs that were several million dollars, reflecting the high capital costs of transit rail cars, track, and stations. The Texas project, which was the largest at $70.4 million, received substantial funding from other sources. This is often the case for large capital investment projects which receive funding from multiple sources, including Federal, State, and local programs.

Service Upgrades/Amenities

This category of CMAQ projects includes strategies to increase transit marketing, provide more widely accessible transit information, improve transit passenger amenities, and create new intermodal connections at transit stations (e.g., improved bus circulation, parking, and interface between bus and rail). Five CMAQ-funded service upgrades/amenities were reviewed in this analysis.

STATE	CMAQ FUNDING	TOTAL COST	PROJECT TITLE	Year Funded	VOC (kg/day)	CO (kg/day)	NOx (kg/day)	PM10 (kg/day)	PM2.5 (kg/day)
Massachusetts	$388,000	$625,000	Fitchburg Intermodal Trans. Center Parking Garage	2002	-14.0	-143.0	-27.0	NR	NR
Missouri	$960,000	$1,200,000	Operation Welcome Aboard Infrastructure (bus shelters)	2004 - 2006	-2.5	NR	-3.4	NR	NR
New York	$160,000	$200,000	Suffolk County Transit Marketing Program	2004	-2.4	-40.7	-2.2	-0.07	-0.03
Ohio	$2,800,000	$3,500,000	Laketran AVL-MDT System	2005	-4.0	-47.0	-13.0	NR	NR
Connecticut	$89,000	$111,000	Rail Utility Construction & Parking Spaces	2007	-6.0	NR	-6.0	NR	-1.0

NR – Values were not reported by the local project sponsor or State DOT in the CMAQ database or other materials for the project.

Increased marketing, provision of more widely accessible transit information, additional customer service, and availability of parking may increase the number of people using public transportation. For instance, Operation Welcome Aboard in Missouri is a passenger amenity project to construct bus shelters at 100 highly utilized stops throughout the Kansas City transit service area. The new facilities will have a coordinated look and feel with the bus fleet and feature valuable route and schedule information. While the project will not expand existing bus routes or create new transit services, the project sponsors estimate that an additional 450 individuals will ride transit each day as a result. The installation of Automatic Vehicle Location (AVL) and Mobile Data Terminal (MDT) systems on Laketran transit vehicles in Ohio is designed to improve the system's paratransit operations, by improving schedule adherence, improving route planning and scheduling, and reducing operating costs; it is estimated to result in a 17.5 percent increase in paratransit ridership.

Congestion/Mobility Benefits

Transit service upgrades and amenities may improve mobility if they make it easier for the public to use public transportation and rely less on their personal vehicles. Since these projects focus on increasing the number of transit riders, they potentially can reduce traffic congestion by reducing the number of personal vehicle trips taken each day. Travel behavior studies have long shown that transit riders respond positively to service improvements that reduce travel or waiting time. Adding more vehicles so as to reduce headways and wait time, or providing routing improvements that reduce travel time or increase reliability are all strategies that can increase ridership. Providing riders with a seat or less crowding can also make the trip more enjoyable, comfortable, and safe, helping to increase the number of transit trips (and reduce the use of SOVs) by encouraging more frequent use by existing riders and attracting individuals who would otherwise drive private vehicles.

Project sponsors for the selected projects reported reductions in vehicle trips ranging from 176 per day (for the Suffolk County Transit Marketing) to 490 per day (for the Fitchburg parking garage at the MART intermodal transportation center). However, since benefits from service upgrades are typically indirect, the projects selected for this study did not assess impacts on delay and speed.

Emissions Benefits

Emissions reduction estimates reported by project sponsors were generally small to moderate. Daily VOC emissions reductions associated with each project range from 2.4 kg to 14.0 kg. Daily NOx emissions reductions associated with each project range from 2.2 kg to 27.0 kg. CO reductions reported by sponsors indicate a range of benefits from 40.7 kg to 143.0 kg emissions reductions each day. CO, PM_{10}, and $PM_{2.5}$ emissions were not reported for some of the projects, but would be expected to drop for each of the selected projects.

Costs

The total public cost of these projects ranged in magnitude from $200,000 to $3,500,000. The non-CMAQ share of project funding ranged from 20 percent to 38 percent of the total project cost.

Transit Vehicle Replacements and Related Infrastructure

Vehicle replacements are designed to reduce the emissions rates of vehicles due to improved technologies or switching to cleaner alternative fuels. While this category is primarily dominated by transit bus purchases, it can also include other public vehicles, such as school buses or government fleets, and related infrastructure, such as fueling stations. Generally, these strategies do not affect congestion levels; instead, they focus primarily on emissions reductions.

Alternative Fuel Vehicles/Fueling Facilities

Projects to purchase alternative fuel vehicles or construct refueling facilities and related other infrastructure are included in this category. Four CMAQ-funded alternative fuel projects were reviewed in this analysis.

STATE	CMAQ FUNDING	TOTAL COST	PROJECT TITLE	Year Funded	VOC (kg/day)	CO (kg/day)	NOx (kg/day)	PM10 (kg/day)	PM2.5 (kg/day)
Maine	$150,000	$1,305,903	Compressed Natural Gas Fueling Station	2002	-2.8	NR	-2.1	NR	NR
Pennsylvania	$5,608,000	$7,010,000	Purchase 12 Alternative Fuel Buses	2002	-3.0	-12.0	-91.0	NR	NR
Connecticut	$688,800	$861,000	CT Clean Fuels Program	2005	-6.8	NR	-12.5	NR	NR
New York	$1,000,000	$1,250,000	Purchase 3 CNG Transit Buses	2007	-1.5	-7.6	-4.3	NR	-1.4

NR – Values were not reported by the local project sponsor or State DOT in the CMAQ database or other materials for the project.

Vehicles that use non-conventional fuels, such as CNG, LNG, electric, or hybrid electric, will reduce emissions while generally having little to no impact on overall VMT. Vehicles that operate using these fuels generally emit fewer pollutants than similar vehicles which run using gasoline or diesel. Other projects provide funding to construct facilities to service, fuel, or provide maintenance for the vehicles in order to encourage their continued use. Alternative fuel vehicle projects provide States and MPOs the opportunity to use high-profile fleets, such as public transit and school districts, to increase public awareness and approval of alternative fuels. This may lead to interest in other fleet operators in switching to alternative fuels.

To encourage alternative fuel vehicle projects to be undertaken in partnership with the private sector, the Transportation Equity Act for the 21st Century contained special provisions for alternative fuel projects that are part of a public-private partnership. For purchase of privately owned vehicles or fleets using alternative fuels, CMAQ funds may be used for only the incremental cost of an alternative fuel vehicle compared to a conventionally-fueled vehicle. Furthermore, if other Federal funds are used for vehicle purchase, such funds should be applied to the incremental cost before CMAQ funds are applied.[27]

Congestion/Mobility Benefits

These strategies are unlikely to reduce congestion since they are not changing transit service in a manner that would be expected to affect ridership. However, bus replacements may increase transit ridership to a

---

[27] See Federal Highway Administration Guidance. *CMAQ and Alternative Fuel Vehicle Projects.* (2005)
http://www.fhwa.dot.gov/environment/cmaqpgs/altfuel/index.htm.

small degree by improving the ease and comfort of transit or improving the reliability of service. These effects are very difficult to determine, and none of the sponsors of the selected projects estimated this effect.

Emissions Benefits

Emissions reductions estimates associated with the replacement of transit vehicles are attributable solely to the lower emissions rates of the new vehicles, not to any effects on transit ridership and diversion of trips from private vehicles. An important consideration with these types of projects is the service life of urban transit buses, which is generally at least 12 years. According to FTA regulation, Federal funds cannot be used to replace vehicles before the end of their useful service life.[28] However, according to EPA guidance for taking credit for emissions reductions, credit can only be taken for the remaining years of service of the older vehicle, not the entire service life of the new vehicle.[29] Consequently, transit vehicle replacement projects will have an immediate emissions benefit when the older vehicle is replaced; however, they likely only have a few years of emissions benefits, over the period of time when the vehicle has reached the end of its service life but might still be continuing in service. The emissions benefits calculations presented in the selected examples only reflect the first year of benefit, and probably should only be assumed for a maximum of a few additional years.

Emissions reductions reported by project sponsors for this set of projects generally indicate the largest emissions reductions from NOx. Daily VOC emissions reductions associated with each project range from 1.5 kg to 6.8 kg. Daily NOx emissions reductions associated with each project range from 2.1 kg to 91.0 kg. CO reductions reported by sponsors indicate a range of benefits from 7.6 kg to 12.0 kg emissions reductions each day. Two project sponsors did not report any CO emissions benefits. $PM_{2.5}$ emissions reductions were reported by one project to be 1.4 kg each day.

Costs

The total public cost of these reviewed projects ranged in magnitude from $861,000 to $7,010,000. The non-CMAQ share of project funding was 20 percent of total project cost for most of the analyzed projects. A natural gas fueling station for public and private fleets operating in the Greater Portland area, Maine, received most of its funding from sources other than CMAQ.

---

[28] See 49 U.S.C. 5309.

[29] This approach is consistent with EPA guidance on diesel engine retrofits ("Diesel Retrofits: Quantifying and Using Their Benefits in SIPs and Conformity - Guidance for State and Local Air and Transportation Agencies", June 2006) and on early retirement of vehicles ("Guidance for the Implementation of Accelerated Retirement of Vehicles Programs", February 1993).

Conventional Bus Replacements

Conventional bus replacement projects replace older diesel buses with new diesel vehicles that emit fewer pollutants. Two bus replacement projects were reviewed in this analysis.

STATE	CMAQ FUNDING	TOTAL COST	PROJECT TITLE	Year Funded	VOC (kg/day)	CO (kg/day)	NOx (kg/day)	PM10 (kg/day)	PM2.5 (kg/day)
Maryland	$5,000,000	$26,500,000	100 Replacement Local Buses	2002	-17.0	NR	-188.9	NR	NR
Ohio	$4,864,440	$6,949,200	61 Replacement Local Buses	2003	-9.6	-35.5	-11.6	NR	NR

NR – Values were not reported by the local project sponsor or State DOT in the CMAQ database or other materials for the project.

The projects included in this category take advantage of improvements in heavy duty diesel technology. Diesel engines today are cleaner and emit fewer pollutants than similar engines more than 10 years ago, so in principle, retiring the older, more-polluting buses will reduce emissions. The new, less polluting vehicles run along existing routes and do not change overall vehicle mileage or service levels, so have no claimed effect on transit ridership. Conventional bus replacement projects have historically comprised a large share of CMAQ funding requests; however, in recent years States and MPOs have opted to replace aging bus fleets with CNG or other alternative fueled vehicles whose emissions rates are even lower than current generation diesel buses.[30] These projects were captured in the Alternative Fuel Vehicles/Fueling Facilities project category.

Congestion/Mobility Benefits

These strategies are unlikely to have congestion or mobility benefits since they are not changing transit service in a manner that would be expected to affect ridership. However, conventional bus replacements may increase transit ridership to a small degree by improving the ease and comfort of transit or improving the reliability of service. These effects are very difficult to determine, and neither of the sponsors of the selected projects estimated this effect.

Emissions Benefits

The emissions benefits from these strategies would be subject to the same caveats applied to alternative fuel vehicle projects. Specifically, if replacement buses are purchased for the purpose of replacing buses that have remaining service life, the emissions credit can only extend to the period of remaining service life of the vehicle being replaced, and with the presumption that the older vehicle will not still be operated.

With these caveats in mind, emissions reported by the sponsors of the example projects indicated daily VOC emissions reductions from 9.6 kg to 17.0 kg. Daily NOx emissions reductions associated with the projects ranged from 11.5 kg to 188.9 kg. CO reductions reported by one project indicated a benefit of 35.5 kg emissions reductions each day.

Costs

CMAQ funding is provided for the capital investment in the new transit vehicles. The total public cost of these projects ranged in magnitude from $6,949,200 to $26,500,000. However, given the expense of bus purchases, CMAQ funds are often used only to supplement FTA funding and are a small share of the overall funding. In the case of the Maryland bus replacement project, more than 80 percent of funding came from sources other than CMAQ.

---

[30] Transportation Research Board. *Special Report 264: The CMAQ Program: Assessing 10 Years of Experience.* 2002.

Dust Mitigation Projects

Road dust reduction strategies are designed to reduce the amount of fugitive dust (PM_{10} and $PM_{2.5}$) that is suspended into the air by tires on roadways. Three CMAQ-funded dust mitigation projects were reviewed in this analysis.

STATE	CMAQ FUNDING	TOTAL COST	PROJECT TITLE	Year Funded	VOC (kg/day)	CO (kg/day)	NOx (kg/day)	PM10 (kg/day)	PM2.5 (kg/day)
California	$174,360	$197,360	Graaf Avenue Paving Project	2004	NR	NR	NR	-143.0	NR
Idaho	$319,600	$319,600	Lincoln Ave Paving Project	2004	NR	NR	NR	-175.5	NR
Idaho	$152,889	$165,000	Purchase of a Liquid De-Icer Truck	2005	NR	NR	NR	-6,292.0	NR

NR – Values were not reported by the local project sponsor or State DOT in the CMAQ database or other materials for the project.

Particles suspended by vehicular movement on paved and unpaved roads are a major contributor to fugitive dust emissions. The origins of this particulate matter differ from the particulate matter that is emitted from vehicles' tailpipes. Exhaust particulate emissions are created from engine combustion while dust mitigation projects control particulate matter originating from the roadway. When vehicles travel along roads, the force of the wheels on the road surface causes the pulverization of surface material. The particles are lifted and dropped from the rolling wheels, and the road surface is exposed to strong air currents in turbulent shear with the surface. The air disturbance behind the vehicle continues to act on the road surface after the vehicle has passed.[31]

Typical dust mitigation projects include paving shoulders, curbs and gutters, roads, and access points. When paving is not feasible, such as for industrial roads with heavy vehicles and/or spillage of material in transport, watering or chemical suppressants may be used. Other CMAQ projects to address the amount of particulate matter released into the air include adding street sweepers, replacing non-certified sweepers with newer vehicles, using new vehicles to increase the frequency of sweeping in existing areas, or using new vehicles to expand the area that is regularly swept. Regular street sweeping on paved roads removes sand and/or other de-icing materials, and other deposition of dirt on roads, reducing the level of road dust.

Congestion/Mobility Benefits

These strategies will have limited, indirect impact on congestion levels, though some benefits may be observed through speed improvements on previously unpaved or icy roads. Since dust mitigation projects are not intended to improve congestion, the sponsors for projects selected for this study did not assess travel impacts.

Emissions Benefits

The quantity of dust emissions from a given segment of road depends on various factors such as whether it is paved or unpaved, precipitation levels, and traffic volumes. Emissions reductions reported by project sponsors at the local level indicated a range of daily PM_{10} emissions reductions from 143.0 to 6,292.2 kg.

Costs

CMAQ funding is usually provided for capital improvements, such as the paving of a road shoulder or purchase of a new street sweeper. The total public cost of the selected projects ranged in magnitude from

---

[31] See U.S. EPA. *AP 42, Fifth Edition*, Volume I, Chapter 13: Miscellaneous Sources. Unpaved Roads. http://www.epa.gov/ttn/chief/ap42/ch13/draft/d13s0202.pdf.

$165,000 to $319,600. The non-CMAQ share of funding ranged from 0 percent to 11 percent of the total cost.

Freight/Intermodal Projects

An intermodal system includes both origins and destinations (for example, ports railheads and warehouses), as well as the links between them (such as roads or rail).[32] Strategies that reduce emissions from the movement of freight and cargo through air quality nonattainment areas are grouped together in the category of freight/intermodal projects. Six CMAQ-funded freight/intermodal projects were reviewed in this analysis.

STATE	CMAQ FUNDING	TOTAL COST	PROJECT TITLE	Year Funded	VOC (kg/day)	CO (kg/day)	NOx (kg/day)	PM10 (kg/day)	PM2.5 (kg/day)
Maine	$283,941	$355,180	South Portland Truck to Rail Intermodal Facility	2000	-0.7	NR	-4.2	NR	NR
Maine	$128,501	$494,098	South Portland – Rail Line Rehab for Freight Shipping	2002	-0.2	NR	-2.0	NR	NR
Pennsylvania	$7,600,000	$9,500,000	Westmoreland Intermodal Freight Facility	2002-2003	NR	-1.9	-13.3	NR	NR
New York	$1,700,000	$9,000,000	Arlington Intermodal Yard	2004	-209.0	-1,712.2	-1,008.8	-37.0	-30.1
Pennsylvania	$10,000,000	$12,500,000	Norfolk Southern Rail Extension and Rehabilitation	2004	-11.5	-64.7	-53.5	NR	NR
Connecticut	$1,409,600	$1,762,000	Freight Rail Construction along Waterfront Street.	2006	-0.5	NR	-18.4	NR	-0.2

NR – Values were not reported by the local project sponsor or State DOT in the CMAQ database or other materials for the project.

Emissions from heavy-duty trucks and large-scale freight facilities can contribute significantly to the overall air pollution in urban areas. Projects that shift movement to a more efficient mode of transport or improve the efficiency of freight transfers between modes will reduce emissions. Some strategies will shift trips from road to rail, reducing emissions and congestion caused by heavy duty vehicles. Other intermodal projects improve the efficiency of transfers between water-borne, truck, and/or rail vehicles. By reducing the amount of time vehicles are required to wait at these transfer stations, idling emissions will be reduced.

Congestion/Mobility Benefits

Reducing the movement of freight by heavy-duty trucks through urban areas can result in congestion relief benefits. Each of the sample projects was designed to reduce truck vehicle travel by shifting freight movement to rail. For instance:

- The sponsors of the Maine projects estimated reductions of up to 2,250 trucks per year by 2006 due to construction of two projects: rail siding as part of an intermodal transfer and rehabilitation and/or replacement of tracks.

- The Westmoreland Intermodal Freight Facility, a project to reduce the amount of freight cargo traveling through downtown Pittsburgh, was estimated to reduce 14 miles of travel for 20,000 truck loads.

---

[32] See Federal Highway Administration Guidance. *CMAQ and Intermodal Freight Transportation.* (2005) http://www.fhwa.dot.gov/environment/cmaqpgs/intermodal/index.htm.

- In New York, capacity improvements to a rail yard were expected to increase rail efficiencies and reduce the movement of freight shipments by truck though the metropolitan New York area. The project sponsors estimated that 10,268 truck trips per day would reduce 6 miles of travel for one segment (Visy Paper Mill – Bayonne Bridge and Transfer Station), and 15,786 truck trips per day would reduce 5 miles for the other segment.

- The Norfolk Southern rail extension and rehabilitation project in Pennsylvania was designed to fund the construction of 5.25 miles and rehabilitation of 7 miles of train track in Indiana County to create a more direct route for delivery of coal. The sponsors estimated the project would reduce 43,478 truck trips per year.

- In Connecticut, the installation of additional railroad track and the associated utility relocations was expected to reduce congestion by shifting an estimated 4,000 truck shipments per year to rail.

The effects on roadway congestion and speeds, however, were not quantified in the analyses provided to the study team.

Emissions Benefits

Emissions reduction estimates reported by project sponsors vary, depending on the modes of transportation affected by the project and the amount of freight that is moved. The Arlington Intermodal Yard in New York, in particular, reported very large emissions reductions, based on assumptions of significant diversions of truck traffic to rail. Most of the calculations do not account for increased railroad diesel emissions, or any congestion or idling associated with transfers between truck and rail.

Costs

Funding under CMAQ has been used to improve efficiency of truck, rail and marine operations, as well as intermodal freight facilities where air quality benefits can be shown. Capital improvements that increase the efficiency of freight movement between truck and rail, for example, as well as up to three years operating assistance for these types of projects, are appropriate for CMAQ funding if emissions reduction can be demonstrated.[33] The total public cost of these projects ranged from $355,180 to $12,500,000. The non-CMAQ share of project funding ranged from 20 percent to 81 percent of the total project cost.

---

[33] 23 USC 149(b)(1), (3). See FHWA Factsheet. "CMAQ and Intermodal Freight Transportation" Available at: http://www.fhwa.dot.gov/environment/cmaqpgs/intermodal/index.htm.

Diesel Emissions Reduction

Diesel emissions reduction strategies are designed to reduce emissions from on-road and off-road diesel engines (e.g., those used in construction equipment, locomotives, marine vessels), and include use of retrofit technologies and idle reduction technologies.

Diesel Engine Retrofits

The term "retrofit" is broadly defined by EPA to include any technology, device, fuel or system that, when applied to an existing diesel vehicle or engine, achieves emissions reductions beyond that required by EPA regulations at the time of a vehicle or engine's certification. Retrofit technologies may include EPA verified emissions control technologies and fuels and CARB-verified emissions control technologies.[34] Seven CMAQ-funded diesel engine retrofit projects were reviewed in this analysis.

STATE	CMAQ FUNDING	TOTAL COST	PROJECT TITLE	Year Funded	VOC (kg/day)	CO (kg/day)	NOx (kg/day)	PM10 (kg/day)	PM2.5 (kg/day)
Maryland	$5,458,000	$23,036,000	142 Bus Engine Upgrades	2001	NR	NR	NR	NR	-34.8
New York	$1,200,000	$1,500,000	WCDOT Diesel Engine Retrofit of 177 Transit Buses	2004	-3.0	-45.9	+14.6[1]	NR	-2.3[2]
Pennsylvania	$1,793,520	$2,242,520	Install 235 Emissions Reduction Devices on Local Buses	2004	-7.3	-111.2	0	NR	NR
Oregon	$49,692	$62,115	Install filters on 9 trash collection vehicles	2005	-1.4	-2.5	0	-0.3	NR
Michigan	$3,360,000	$4,200,000	3 Locomotive Diesel Engine Retrofits	2007	-10.0	NR	-132.1	NR	-3.7
New York	$424,000	$530,000	Diesel Engine Retrofits of 53 County Vehicles	2007	-0.4	-1.7	0	-0.2	-0.1
New York	$1,368,000	$1,710,000	Rockland County retrofit of on-road diesel vehicles	2007	-140.3	-969.9	0	-138.4	-125.9

NR – Values were not reported by the local project sponsor or State DOT in the CMAQ database or other materials for the project.

[1] Project sponsor calculated an increase in NOx emissions based on data from retrofits and existing emissions from tailpipe testing. The project sponsors noted that the increase in NOx emissions is highly unusual and it is widely accepted that these retrofits have no impact on NOx.

[2] Project sponsor did not report PM reductions in CMAQ database but included information on emissions rates in project backup information to enable calculation.

Diesel engine retrofits are typically aimed especially at reducing particulate matter from heavy-duty diesel engines, as well as other pollutants. Verified technologies purchased and installed through these projects included the Englehard DPX soot filter and bus engine overhauls. As with all CMAQ projects,

---

[34] A list of the EPA verified technologies can be accessed at: www.epa.gov/otaq/retrofit/verif-list.htm.

retrofitted vehicles must operate predominantly within or in close proximity to nonattainment or maintenance areas.[35]

CMAQ-funded diesel retrofit projects include a wide range of measures to reduce diesel emissions by retrofitting vehicles/equipment with new or improved emissions control equipment, upgrading engines, replacing older engines with newer/cleaner engines, and using cleaner fuels. The selected projects included installation of retrofit devices on transit buses, trash collection vehicles, a range of county-owned vehicles, and locomotives.

Congestion/Mobility Benefits

Diesel engine retrofits are not designed to provide congestion benefits and do not affect travel.

Emissions Benefits

Project sponsors reported that daily VOC emissions reductions ranged from 0.4 kg to 140.3 kg. One project sponsor did not report any VOC emissions effects. Most of the selected projects were not expected to reduce NOx (i.e., particulate filters, such as the Englehard DPX soot filter, reduce emissions of PM, VOC, and CO, but not NOx), according to EPA's Diesel Retrofit Technology Verification.[36] However, the Michigan locomotive repowering project was estimated to reduce a substantial amount of NOx, due to lower fuel use and an 86 percent estimated reduction in ozone precursors. A reported increase in NOx emissions from a diesel engine retrofit project is highly unusual. The project sponsor for the New York project that reported noted that it is widely accepted that these retrofits have no impact on NOx; however, an increase was reported based on data from the retrofit manufacturers and emissions from tailpipe testing of the subject vehicles.

PM reductions of 0.1 to 138.4 kg/day were reported for the sample retrofit projects. In the case of the Pennsylvania bus retrofit project, no PM emissions reductions were reported; however, EPA reports a 60 percent reduction in PM emissions associated with the Englehard DPX soot filter. The CO reductions reported by sponsors indicate a range of benefits from 1.7 kg to 969.9 kg emissions reductions each day. Two projects did not report any CO emissions benefits.

Costs

CMAQ funding is usually provided for equipment such as new diesel engines, truck stop electrification infrastructure, or purchase of retrofit devices. Funding may also be provided to offset a portion of the cost of installation or operation of a regional retrofit program. The total public cost of these projects range in magnitude from $530,000 to $23,036,000, depending on the number of vehicles to be retrofitted and the type of device used. The non-CMAQ share of project funding ranged from 20 percent to 76 percent of the total project cost.

---

[35] 23 USC 149(b)-(c). See FHWA, 2003. "Eligibility of Freight Projects and Diesel Engine Retrofit Programs" Memorandum. http://www.fhwa.dot.gov/environment/cmaqpgs/retrom.htm.

[36] For a listing of verified retrofit technologies and their emissions reductions, see: http://www.epa.gov/oms/retrofit/verif-list.htm.

Truck Idle Reduction

Unnecessary idling often occurs when trucks wait for extended periods of time to load or unload materials or supplies, or when equipment is left on overnight when it is not being used. Idle reduction strategies eliminate this unnecessary idling by heavy duty vehicles which can save fuel, prolong engine life, and reduce emissions. There are several technologies available to address idling. Some of these technologies are mobile and attach onto the truck (mobile Auxiliary Power Units (APUs)), and provide air conditioning, heat, and electrical power to operate auxiliaries such as a microwave. Another technology involves electrifying truck parking spaces (stationary Truck Stop Electrification (TSE)) with or without modifying the truck. This involves power from the electrical grid providing energy to operate stationary equipment or on-board truck equipment to provide cab heating, cooling, and other needs.[37] Three truck idling reduction projects were reviewed in this analysis.

STATE	CMAQ FUNDING	TOTAL COST	PROJECT TITLE	Year Funded	VOC (kg/day)	CO (kg/day)	NOx (kg/day)	PM10 (kg/day)	PM2.5 (kg/day)
Tennessee	$1,000,000	$1,000,000	100 Auxiliary Power Units	2003	-4.5	NR	-60.4	NR	NR
Kentucky	$500,000	$835,000	50 Auxiliary Power Units	2005	-6.7	- 46.7	-110.2	NR	NR
Tennessee	$788,240	$985,300	59 Auxiliary Power Units	2006	NR	NR	-79.7	NR	2.2

NR – Values were not reported by the local project sponsor or State DOT in the CMAQ database or other materials for the project.

For long haul trucks, the truck driver must have 10 hours off duty after driving 11 hours.[38] Surveys have found that 70 to 80 percent of truck drivers say the need for heating or air conditioning is the main reason they idle their trucks while off duty. They also cite the need to operate on-board electrical appliances, such as a television or refrigerator, and to ensure the engine block, fuel, and oil remain warm. Long duration truck idling occurs at truck stops, travel centers, distribution hubs, airports, borders, ports, and roadsides.[39]

Congestion/Mobility Benefits

Truck stop idle reduction projects are not designed to provide congestion benefits and do not affect travel.

Emissions Benefits

Truck stop idle reduction projects are designed primarily to reduce NOx and PM emissions. While some project sponsors in the past estimated reductions in VOC and CO using MOBILE idle emissions factors, EPA's "Guidance for Quantifying and Using Long Duration Truck Idling Emission Reductions in State Implementation Plans and Transportation Conformity" (January 2004) provides long-duration idling emissions factors only for NOx and PM. At the time the emissions calculations were conducted for the 2003 Tennessee project and the 2005 Kentucky project, this guidance had not yet been available or used.

The three projects reported NOx emissions reductions estimates of 60.4 to 110.2 kg/day.[40] Only the 2006 Tennessee projects have estimated $PM_{2.5}$ emissions reductions of 2.2 kg/day. Using long-duration truck

---

[37] See Federal Highway Administration Guidance. *CMAQ and Idle Reduction Technologies.* (2005) http://www.fhwa.dot.gov/environment/cmaqpgs/idlereduct/index.htm.

[38] See 49 CFR, Part 395. For additional information: http://www.fmcsa.dot.gov/rules-regulations/topics/hos/hos-2005.htm.

[39] See Federal Highway Administration Guidance. *CMAQ and Idle Reduction Technologies.* (2005).

[40] The 2003 Tennessee project had an estimated 60.4 kg/day in NOx based on MOBILE emissions factors; using long-duration idle emissions factors available from the EPA guidance, released in 2004, the project would reduce 135.0 kg/day.

idling emissions factors currently available from EPA, the 2003 Tennessee and 2005 Kentucky projects would have had PM emissions reductions of 3.7 kg/day and 1.8 kg/day, respectively.

Costs

CMAQ funding is usually provided for truck stop electrification infrastructure and equipment. The total public cost of these projects ranged in magnitude from $835,000 to $1,000,000.

4. PROJECT ANALYSIS AND SELECTION PRACTICES THAT SUPPORT EFFECTIVENESS

In the previous section of this report, data gathered from the national CMAQ database and local project sponsors were presented to document reported congestion and emissions benefits, as well as characteristics of the broad types of strategies. This section focuses on using information from the selected set of projects to assess the projects' air quality cost-effectiveness and to examine how some areas are using this type of information for program prioritization and decision making.

This section is divided into two parts. First, a discussion of the cost-effectiveness of the selected projects at reducing emissions of each of the primary pollutants – VOC, CO, NOx, and PM_{10} and $PM_{2.5}$ – is provided. In order to calculate air quality cost-effectiveness in a way that allows appropriate comparisons, project costs and emissions effects have been recalculated to fill in gaps in reported emissions reductions and to "normalize" the results to a common year, 2008.

Second, initial observations on good practices that States and MPOs have used to analyze, prioritize, and select CMAQ projects; including use of cost-effectiveness analysis and consideration of other factors are provided. Phase II of this evaluation and assessment study further expand upon this information through development of case studies of specific locations to understand State DOT and MPO practices and to help enhance the effectiveness of the program.

Emissions Reduction Cost-Effectiveness

The Role of Cost-Effectiveness Assessment

Understanding the cost-effectiveness of CMAQ projects should be an important consideration in project selection decisions at the State and local level. SAFETEA-LU directs that States and MPOs give priority to "diesel retrofit projects and. . .other cost-effective emission reduction activities, taking into consideration air quality and health effects" and to "cost-effective congestion mitigation activities that provide air quality benefits."[41] Moreover, States and MPOs, as good stewards of public dollars, will maximize the value of their investment of CMAQ funds by targeting it toward projects that provide the most benefit per dollar. Indeed, conducting a cost-effectiveness assessment provides States and MPOs with the ability to stretch limited transportation funding resources across a wide range of projects that demonstrate congestion, energy, environment, air quality, and mobility benefits.

Given the role of the CMAQ program as a key funding source to help transportation agencies meet air quality goals consistent with attainment of regional air quality plans, cost-effectiveness at reducing air pollutant emissions is often considered an important metric of CMAQ program effectiveness. At the same time, it is important to recognize that the benefits of the CMAQ program go well beyond emissions reduction, and States and MPOs often take into account these other considerations in making project selection decisions. In this study, cost-effectiveness for the 67 sample projects was calculated in regard to emissions reductions alone due to the availability of information on emissions reduction estimates for a wide variety of CMAQ-funded transportation projects. However, emissions reduction cost-effectiveness may not be the only measure of cost-effectiveness for a project, just as air quality is not the only benefit that may be considered in project selection.

In many urban areas and states with severe traffic congestion problems, a project's cost-effectiveness at alleviating traffic congestion will often be an important consideration. A project that reduces traffic congestion in a targeted corridor may be viewed as more beneficial than another project that reduces the same level or more emissions but does not provide congestion relief benefits. These congestion relief

---

[41] SAFETEA-LU 1808(d).

benefits are difficult to quantify using a standard metric (such as hours of traveler delay reduced) across all projects, based on the complexities of modeling and assessing these impacts, particularly for small projects.

CMAQ projects generate a wide range of other benefits, which may also be important factors in project selection. These benefits include, among others, enhancing mobility and access, creating more reliable travel times and transit services, encouraging physical activity, reducing greenhouse gas emissions, creating better connections between transportation and land use, and fostering a more multi-modal transportation system. Most of these benefits are difficult, if not impossible, to quantify in a standard metric, and thus are not usually considered in a cost-effectiveness framework. However, these benefits may be very important in the context of regional transportation goals. The flexibility inherent in the CMAQ program allows local areas to determine their own procedures and criteria for project assessment. States and MPOs are using a suite of evaluation criteria, including air quality and energy conservation benefits, local cost participation share, and intermodal, multi-modal, and social mobility concerns, to ensure all are being addressed in regional transportation planning and programming.[42]

Methodology for Analyzing Emissions Reduction Cost-Effectiveness of the Selected Projects

The study team calculated cost-effectiveness of the sample projects with respect to reductions of VOC, NOx, CO, PM_{10}, and $PM_{2.5}$. Cost-effectiveness figures were developed for each pollutant independently, rather than as a composite figure. This was done for two primary reasons: 1) At the national level, it is difficult to determine the most appropriate means of weighting each pollutant, given that some pollutants are of more concern in some parts of the country than others. 2) Some strategies are targeted toward reducing individual pollutants, such as dust mitigation projects, which focus on PM_{10} reduction. Lumping together the reduction of a full set of pollutants, therefore, would not show how different types of strategies can be more or less effective at reducing different pollutants.

In order to increase the comparability of emissions and cost figures across the sample projects, the study team recalculated project costs and emissions effects. Recalculations were conducted largely because for many projects, data were missing on specific pollutants – commonly CO and PM. In addition, the projects were implemented in a wide range of different locations, at different times, and emissions benefits were reported for different years. Since the U.S. vehicle fleet is on average, much cleaner today than it was 10 years ago, a project that eliminates a mile of travel will have less emissions reduction benefit in 2010 than the same project in 2000. Consequently, it was useful to standardize the emissions effects to a common year using a standard set of default emissions factors for purposes of analyzing cost-effectiveness across the selected projects. Project costs, including operating and capital costs, were also adjusted to reflect constant 2008 dollars to enable better comparisons. Costs were standardized using the Consumer Price Index (CPI) which may result in reduced cost-effectiveness for multi-year projects.

The following discussion provides more detail on the calculation procedures used in this study. The procedures for "normalizing" emissions and costs are described below first. This is followed by a description of the general steps in conducting cost-effectiveness analysis, which could be used for any calculations of project cost-effectiveness, including those conducted at the State or local level. In fact, the study team found that a number of State DOTs and MPOs were using the same basic approach to calculate cost effectiveness for their proposed projects.

PROCEDURES TAKEN TO ENSURE COMPARABILITY OF COST AND EMISSIONS DATA FROM SAMPLE PROJECTS

The "normalization" procedures used to standardize the projects in this study included three main steps.

---

[42] Integrating Air Quality and Transportation Planning: A Compendium of Workshop Summaries for Regional Councils and metropolitan planning organizations. 2001-2005. Available at: www.narc.org/uploads/File/01Workshop_Summaries_2005_Edit.pdf.

1. Establish baseline running and trip start emissions factors for 2008 across multiple pollutants.

2. Recalculate emissions reductions using standardized calculation methodologies and 2008 emissions factors.

3. Recalculate project costs in 2008 dollars using the Consumer Price Index (CPI).

These steps are described below.

1) Develop "Normalized" Emissions Factors. In order to improve comparability of results, a common set of emissions factors for CO, NOx, VOC, PM_{10}, and $PM_{2.5}$ emissions was developed using MOBILE6.2 for analysis year 2008. MOBILE is EPA's approved model for estimating pollution from highway vehicles. The model calculates emission factors (in grams per vehicle-mile) for a variety of pollutants from passenger cars, motorcycles, light- and heavy-duty trucks. Some of the emissions factors are based on testing of tens of thousands of vehicles and account for changes in vehicle emission standards over time, changes in vehicle populations and activity levels, and variation in local conditions such as temperature, humidity and fuel quality.[43]

In the data collected from State DOTs and MPOs, some projects' emissions effects had been calculated with an earlier version of the model, MOBILE5a, or EMFAC, the California emissions model. In this analysis, the 2008 emissions factors from MOBILE6.2 were applied to all the selected projects, where feasible, reducing differences due to the time the project was implemented, local weather and vehicle fleet mix, and/or region-specific modeling assumptions. For additional information on the assumptions and inputs used to develop the normalized emissions factors, please see Appendix B. It should be noted that while this normalization was helpful for purposes of this study, State DOTs and MPOs should not apply this procedure in their own project assessments. They should use the best available data at the local level to develop appropriate emissions factors for conditions in their area.

Emissions factors for some types of CMAQ projects, such as diesel engine retrofits, bus replacement, and dust mitigation projects, were not standardized. These technologies vary widely in their ability to achieve emissions reductions and depend on specific local conditions (e.g., road dust levels depend on precipitation and silt loadings). Consequently, the emissions factors reported by project sponsors, or the most recent EPA certification data for retrofits, were used in the calculation of cost-effectiveness.

2) Recalculate Emissions Reductions, as appropriate. Using the normalized emissions factors, emissions reductions were then recalculated (as kg/day). This generally was done using the project's reported travel impacts (e.g., VMT reductions, speed changes), using the methodologies that the project sponsors had used.

3) Adjust Total Public Costs to Constant Dollars. CMAQ and non-CMAQ project costs reported by local sponsors were converted to a 2008 base by using the average annual Consumer Price Index published monthly by the U.S. Bureau of Labor Statistics.[44] CPI values for respective years in relation to 2008 are shown in Table 3, along with the corresponding adjustment factor. The CPI for January 2008 is 211.08. To convert dollar values in one year to constant dollars in a second year, multiply the first-year dollar value by a factor whose numerator is the average annual CPI of the second year and whose denominator is the average annual CPI of the first year. For instance, to convert $10,000 in 2000 dollars to 2008 constant dollars, multiply $10,000 by the average annual CPI in 2008 divided by the average annual CPI in 2000:

---

[43] U.S. Environmental Protection Agency. Office of Transportation and Air Quality. *Description and History of the MOBILE Model.* (2004) http://www.epa.gov/OMS/mobile.htm.

[44] U.S. Department Of Labor, Bureau of Labor Statistics. Consumer Price Index for All Urban Consumers - (CPI-U), U.S. city average, All items. *ftp://ftp.bls.gov/pub/special.requests/cpi/cpiai.txt.*

$10,000 * (211.1 / 172.2) = $10,000 * 1.226 = $12,259 in 2008 constant dollars.

Table 3. Consumer Price Index (CPI) Factors.

Year	CPI	Factor	Year	CPI	Factor
1999	166.6	1.267	2004	188.9	1.118
2000	172.2	1.226	2005	195.3	1.081
2001	177.1	1.192	2006	201.6	1.047
2002	179.9	1.173	2007	207.3	1.018
2003	184.0	1.147	2008	211.1	1.000

CALCULATING EMISSIONS REDUCTION COST-EFFECTIVENESS USING THE NORMALIZED RESULTS

Once normalized emissions figures and costs were developed, cost-effectiveness at reducing each type of emissions was calculated using a standard approach, as listed in the three steps below:

1) Calculate average annual emissions reduction;
2) Calculate annualized cost of the project;
3) Divide the annualized cost of the project by the annual emissions reduction.

The cost figures used in the calculations represent the total public costs associated with implementing a project. This includes funding from the CMAQ program for capital and operating costs, as well as any other Federal, State, or local sources. Some individual projects were funded over multiple years or multiple States, and so the cost used in the calculations reflected all of these components. In some cases, CMAQ only paid for a small portion of a project's total costs. Some State and MPO analyses that the study team reviewed involved calculations of both overall project cost-effectiveness (based on the full costs of the project) and cost-effectiveness associated with CMAQ dollars alone (not including other Federal, state, and local funding sources). However, the results presented in this study reflect full project cost-effectiveness at reducing emissions using total public funds. The measure of cost-effectiveness reported in this study is dollars per ton (although it can also be reported as dollars per kg, or another similar metric). The steps in this process are described below.

1) Convert to Annualized Emissions Reductions. Emissions reductions per day should be converted into annualized values as kg per year (which the study team converted to tons per year). Although the national CMAQ database reports emissions reductions in kg per day, in most cases the emissions benefits are not realized on all 365 days of the year, but are restricted to only work travel/weekdays or to a smaller number of days when the program is in effect (e.g., on ozone exceedance days, days when bicycling is considered most feasible, or days of application of de-icing chemicals). In most of the project examples, the project sponsor included an estimate of the number of days during which the strategy would be effective. In these situations, the local figure was used. However, in instances where it was not provided, the following standard scaling factors were used:

- For projects that affect only peak period or weekday commuter travel, daily effects were multiplied by 250.

- For projects that affect all traffic, daily effects were multiplied by 365.

2) Calculate Annualized Cost of the Project. To calculate the annualized cost of the project, two pieces of information are needed:

- Project funding – The total project funding cost is needed. This should include funding from the CMAQ program, as well as any other Federal, state, or local sources.

- Capital recover factor (CRF) – The capital recovery factor is used to determine how to annualize funding dollars over the life of the project, assuming that projects with service lives beyond 1 or 2 years represent an opportunity cost in the use of those public resources equal to the value of those resources if invested for the same time period at a societal rate of interest. The capital recovery factor is calculated using the following equation:

$$CRF = \frac{(1 + i)^n \times (i)}{(1 + i)^n - 1}$$

Where i = discount rate (as a decimal fraction)
    n = project life (in years)

The annualized cost of the project is calculated by multiplying total project funding by the CRF, as shown in the following equation:

$$Annualized\ cost = Project\ funding \times CRF$$

The discount rate reflects the rate at which society (taxpayers) values future benefits in terms of resources that it must give up now. As a result, it "discounts", or places a lower value on, future benefits from the investment compared to current benefits. A lower discount rate increases the effective value of future benefits (emissions reductions) by lowering the annualized cost used in the comparison.[45] A 7 percent discount rate was used in this analysis, which is the value used by the Federal Transit Administration (FTA) in its New Starts program, and is the rate recommended by the Office of Management and Budget (OMB) for Federal investment analysis.

"Project life" represents the period of time over which a project remains effective at reducing emissions and congestion levels, and varies by the type of project. For example, a standard transit bus is expected to provide service for 12 years, whereas the service life of a vanpool vehicle may be only 5 years. For some projects, effects last for many years; in other cases, the effects continue only for the length of time when direct funding is provided. Individual project life periods, determined by the specific circumstances of each project and local jurisdiction, were sometimes reported by project sponsors, and these were typically used in the calculations. However, some general rules are provided in Table 4, based on a review of project life periods used by other sources, and these were generally used where no other data were provided.[46]

---

[45] For example, imagine two projects, each reducing 1 pound of emissions per year. The first project costs $1,000 and has 5 years of effects; the other project costs $2,800 but has 20 years of effects. At a 7 percent discount rate, the first project appears more cost effective, while with a 5 percent discount rate, the second project appears more cost effective.

[46] See: California Air Resource Board, "Methods to Find the Cost-Effectiveness of Funding Air Quality Projects," 1999; Birmingham Regional Planning Commission, "A Guide for Estimating the Emissions Effects and Cost-Effectiveness of Projects Proposed for CMAQ Funding," 2002; Maricopa Association of Governments, "Methodology for Evaluating Congestion Mitigation and Air Quality Improvement Projects"; U.S. Environmental Protection Agency, "Summary Review of Cost and Emissions Information for 24 Congestion Mitigation and Air Quality Improvement Projects," 1999; Transportation Research Board Special Report 264, "The Congestion Mitigation and Air Quality Improvement Program: Assessing 10 Years of Experience."

Table 4. Project Life Periods Used for Evaluating Projects.

CATEGORY	SUBCATEGORY	PROJECT LIFE EXPECTANCY (YEARS)
Traffic Flow Improvements	Traffic Signalization	10
	Freeway Management	10
	High-Occupancy Vehicle Lanes	20
Shared Ride Programs	Regional Ridesharing	1 to 2
	Vanpool Programs - ongoing assistance - purchase of vans	1 to 2 5
	Park-and-Ride lots	12
Travel Demand Management	Regional Approaches/Employer Trip Reduction programs	1 to 2
Bicycle/Pedestrian Facilities	-	15
Transit Improvements	New Bus Services - purchase of new buses - operations	12 1
	New Rail Services - Railcars - Track/stations	20 30
	Service Upgrades - Amenities - Bus shelters, etc.	2 10
Technology Improvements (primarily transit)	Conventional Bus Replacements and Alternative Fuel Buses (assumed remaining life of vehicles)	4
Dust Mitigation Projects	-	20
Freight/Intermodal Projects	-	20
Engine Retrofit Technologies	Diesel Engine Retrofits	Varies
	Truck Stop Electrification	10

3) Calculate Cost-effectiveness. Once air pollutant emissions and costs were standardized into annualized values, a cost-effectiveness calculation was determined for each project sample. Cost effectiveness is calculated using the following equation:

$$\text{Cost Effectiveness (\$/ton)} = (\text{Annualized Cost}) / (\text{Annual Emissions Reduction})$$

A project is more cost effective when it achieves its results at the lowest possible cost. For each project, cost effectiveness was calculated according to the estimated reductions of VOC, NOx, CO, PM_{10}, and $PM_{2.5}$ emissions individually.

Results

The results of the cost-effectiveness analysis for each pollutant are presented below, in Tables 7 and 8, which summarize the minimum and maximum cost-effectiveness figures for individual projects studied within each category and subcategory. Given the small number of projects studied, the median value has not been provided.

In examining emissions reductions by individual pollutant, it is important for State DOTs and MPOs to consider the specific air quality issues that are faced in their areas. Moreover, the health effects, emissions inventories, and control sources for each pollutant are also different. For instance, transportation sources produce significantly more CO than PM; correspondingly, reducing a ton of PM often costs more than reducing a ton of CO. The benefits of reducing a ton of PM may also be more valuable, based on health studies showing the significant effects of PM on human health.

Across the project categories, some patterns emerge, although the results are limited due to the small number of projects studied, and cannot be used to determine statistically significant median cost-effectiveness values or confidence intervals. The projects profiled in this study are intended to be illustrative of typical projects funded through the CMAQ program, but do not represent a statistical sampling of the CMAQ database. The largest sample category size in this study is seven diesel engine retrofit projects. It is important to note that these figures are not directly comparable to the results from some other studies, such as the TRB study on the CMAQ program, cited in the Appendix to the CMAQ Interim Guidance.[47]

---

[47] For instance, in the TRB study, emissions benefits for NOx and VOC were combined and weighted, resulting in a composite cost-effectiveness figure, whereas this study presents separate figures for each pollutant. Moreover, in the TRB study, emissions benefits expected to occur in the future were "discounted", but for this study, all emissions benefits were counted equally. The inherent economic logic of discounting presumes that short-term benefits are preferable to benefits in the long-term. This study, however, values a ton of emissions reduction in year 1 as equivalent to a ton of emissions reduction in year 10. This approach makes the reporting of dollars per ton reduced more intuitive when reporting emissions reductions for individual pollutants. It also is consistent with the fact that emissions reductions in a nonattainment area need to be continued into the future in order for compliance with the ambient air quality standards.

Table 5. VOC, NOx, and CO Cost-Effectiveness of Selected CMAQ Projects by Strategy.

Category	No. Projects	VOC ($/ton)		NOx ($/ton)		CO ($/ton)	
		Low	*High*	*Low*	*High*	*Low*	*High*
Traffic Flow Improvements							
Traffic Signalization	7	$2,000	$5.6 M	$5,000	+	$500	+
Freeway Management	4	$1,000	$98,000	$10,000	+	$2,000	+
High-Occupancy Vehicle Lanes	1	$18.9 M		$40.5 M		$1.3 M	
Shared Ride Programs							
Regional Ridesharing	3	$86,000	$494,000	$78,000	$440,000	$7,000	$39,000
Vanpool Programs	4	$34,000	$158,000	$29,000	$160,000	$3,000	$13,000
Park-and-Ride Lots	5	$14,000	$8.5 M	$12,000	$4.9 M	$1,000	$384,000
Travel Demand Management	4	$16,000	$2.9 M	$15,000	$2.9 M	$1,000	$223,000
Bicycle/Pedestrian Facilities	4	$551,000	$6.0 M	$667,000	$7.4 M	$46,000	$453,000
Transit Improvements							
New Bus Services	3	$130,000	$1.5 M	$222,00	$1.4 M	$9,000	$15,000
New Rail Services	3	$88,000	$416,000	$89,000	$380,000	$7,000	$33,000
Service Upgrades/Amenities	5	$11,000	$1.5 M	$7,000	$1.5 M	$1,000	$116,000
Bus Replacements/Technologies							
Conventional Bus Replacements	2	$852,000	$1.5 M	$134,000	$231,000	$706,000	
Alternative Vehicles/Fueling Facilities	4	$152,000	$2.9 M	$82,000	$316,000	$124,000	$734,000
Dust Mitigation Projects	3	--	--	--	--	--	--
Freight/Intermodal Projects	6	$37,000	$424.2 M	$2,000	$213,000	$7,000	$3.7 M
Diesel Emissions Reduction							
Diesel Engine Retrofits	7	$7,000	$677,000		$21,000	$1,000	$174,000
Truck Idle Reduction	3	--	--	$2,900	$4,600	$6,800	

NOTE: Cost-effectiveness calculations noted with a plus sign (+) indicate that project(s) in the category showed an increase in the pollutant of concern. Projects with (--) indicate categories where a cost effectiveness calculation was not applicable due to zero pollution reduced.

Table 6. PM Cost-Effectiveness of Selected CMAQ Projects by Strategy.

Category	No. Projects	PM$_{10}$ ($/ton)		PM$_{2.5}$ ($/ton)	
		Low	*High*	*Low*	*High*
Traffic Flow Improvements					
Traffic Signalization	7	$287,000	$68.9 M	$442,000	$106.2 M
Freeway Management	4	$279,000	$15.7 M	$430,000	$135.9 M
High-Occupancy Vehicle Lanes	1	--	--	--	--
Shared Ride Programs					
Regional Ridesharing	3	$2.0 M	$11.1 M	$4.2 M	$24.1 M
Vanpool Programs	4	$695,000	$3.8 M	$1.5 M	$8.3 M
Park-and-Ride Lots	5	$285,000	$128.2 M	$616,000	$277.5 M
Travel Demand Management	4	$390,000	$79.8 M	$845,000	$172.9 M
Bicycle/Pedestrian Facilities	4	$22.8 M	$259.6 M	$49.4 M	$562.1 M
Transit Improvements					
New Bus Services	3	$6.1 M	$6.1 M	$13.3 M	(+)
New Rail Services	3	$2.3 M	$9.7 M	$5.0 M	$21.2 M
Service Upgrades/Amenities	5	$184,000	$41.6 M	$398,000	$90.1 M
Bus Replacements/Technologies					
Conventional Bus Replacements	2	--	--	--	--
Alternative Vehicles/Fueling Facilities	4	--	--		$676,000
Dust Mitigation Projects	3	$15	$700	--	--
Freight/Intermodal Projects	6	$66,000	$10.8 M	$80,000	$13.2 M
Diesel Emissions Reduction					
Diesel Engine Retrofits	7	$7,000	$1.7 M	$8,000	$2.1 M
Truck Idle Reduction	3	$110,300	$173,600	$110,300	$173,600

NOTE: Cost-effectiveness calculations noted with a plus sign (+) indicate that project(s) in the category showed an increase in the pollutant of concern. Projects with (--) indicate categories where a cost effectiveness calculation was not applicable due to zero pollution reduced. One figure reported between the high and low categories indicates that only one project reported emissions effects for that pollutant.

As seen in the tables, a high level of variability is found in the results for each individual category of projects, indicating that local context and project-specific factors are an important determinant of cost-effectiveness. The range of estimated figures for air quality cost-effectiveness within individual categories is very large, with some individual projects showing very strong cost-effectiveness for certain pollutants, while others clearly appear to have lower cost-effectiveness for certain pollutants, as indicated by costs of several million dollars per ton.

This finding seems to indicate that some projects are better suited for reducing certain pollutants and likely were selected for reasons other than emissions reductions (e.g., congestion mitigation, social effects). Indeed, while air quality cost-effectiveness is an important aspect of transportation agencies' project selection, these other benefits can have significant impacts on overall urban mobility, livability, and sustainability initiatives.

Observations regarding the various categories of projects are noted below.

TRAFFIC FLOW, SHARED RIDE, AND DEMAND MANAGEMENT PROJECTS.

Some traffic flow improvements and projects that target reductions in single-occupancy vehicle travel – such as shared ride and travel demand management programs – were very cost-effective in reducing the ozone precursors, VOC and NOx, as well as CO. Due to the relatively limited contribution of personal motor vehicles to PM, in comparison to VOC, NO_X, and CO, none of these strategies appeared to be very cost-effective at reducing PM. Moreover, the MOBILE6 model used to generate the emissions changes for this analysis does not take into account the impact of changes in vehicle speeds on PM emissions levels.[48] Therefore, the PM emissions reductions reported from traffic flow projects in this analysis were only calculated based on reductions in vehicle idling due to reductions in incident-based or intersection delay. These projects often have important non-emissions benefits, including travel time savings, reductions in greenhouse gases, and supporting increased non-motorized travel.

TRANSIT AND TECHNOLOGY/FUELS PROGRAMS

Transit improvements that target reductions in motor vehicle travel, such as new rail or bus services and service upgrades/amenities, appear to offer the potential for relatively high cost-effectiveness at reducing VOC, NOx, and CO emissions, but fared poorly in reducing PM. Overall, bus replacement projects fared poorly in cost-effectiveness at emissions reduction. The costs are used to purchase vehicles that will last 12+ years in service, but emissions benefits can only be credited for a limited number of years, not the full service life of the new bus.

DUST MITIGATION

Projects focused on dust mitigation offered some of the most effective means measured in this study for reducing PM_{10} and $PM_{2.5}$ emissions in locations where they were practical. These projects, including paving unpaved roads and application of deicing chemicals to reduce sand application, achieved substantial reductions in particulate matter (in the form of wind-blown dust) for far less public resources than other types of project categories.

DIESEL EMISSIONS REDUCTION AND FREIGHT/INTERMODAL PROJECTS

Diesel retrofits, truck idle reduction, and freight/intermodal projects are categories of projects that have received increased emphasis in recent years. These categories had some of the most cost-effective projects within the reviewed projects at reducing both ozone precursors and particulate matter. However, there was a very large range, with some projects fairing poorly when focusing solely on the cost-effectiveness of emissions reductions. This may be due in part to the fact that different retrofit technologies target different pollutants. For instance, one retrofit project showed high cost-effectiveness at reducing NOx,

---

[48] See Preamble to 40 CFR Part 93 for additional discussion of EPA's conformity decisions.

whereas some retrofits showed no impact on NOx.[49] Use of idling reduction technology to reduce long-duration truck idling showed the best cost-effectiveness at reducing NOx emissions.

Examples of Good Practices

States use a variety of processes and procedures to identify, select, and evaluate projects for inclusion in the CMAQ program. Drawing on the observations and results of the project analysis and information from State and local project sponsors, the following sections provide examples to illustrate the range of approaches taken by States and MPOs. Examples of good approaches identified through this research include:

- Use of standardized emissions calculation methodologies and tools in order to help ensure validity and comparability of emissions reduction estimates;
- A documented, transparent project prioritization/selection process, including consultation of States and MPOs with State and local air quality agencies; and
- Collection of post-project data to determine whether projected impacts were achieved.

These practices are discussed below.

Standardized Tools or Emission Calculation Methods

State and local transportation and air quality agencies have the flexibility to conduct CMAQ project air quality analyses with different analytical approaches. While FHWA does not specify a single set of methods for use in CMAQ emissions estimation, every effort should be taken to ensure that determinations of air quality benefits are credible and based on reproducible and logical analytical procedures.

USE OF ACCEPTED EMISSIONS CALCULATION APPROACHES

An important first step in making decisions is to base those decisions on appropriate methodologies and reasonable assumptions. There are several online resources and published guides available to State and local transportation practitioners that describe the modeling tools and other methods that can be used to assess the emissions benefits of projects applying for CMAQ funds.

The most recent and comprehensives of these is a guidebook, *Multi-pollutant Emissions Benefits of Transportation Strategies* (2006). This compendium includes sketch planning methods for 35 different categories of transportation strategies, based on a review of many guidance documents and analytical tools. The report includes calculations of emissions impacts for sample projects, based on real project examples, and identifies EPA and FHWA guidance documents that should be referenced. It also reports on the direction of emissions impacts (increase, decrease, neutral or uncertain) that are typically expected for each transportation strategy on the following seven pollutants: CO, PM_{10}, $PM_{2.5}$, NOx, VOCs, SOx, and NH_3. The report is available at: www.fhwa.dot.gov/environment/conformity/mpe_benefits.

The report *A Sampling of Emissions Analysis Techniques for Transportation Control Measures* (2000) includes a brief overview of 19 methods which include pre-packaged and customizable software tools as well as worksheets or other procedures for calculating benefits. They collectively address a wide range of potential CMAQ projects, including travel demand management, traffic flow improvements, and vehicle

---

[49] For comparison purposes, EPA's document, "The Cost-Effectiveness of Heavy-Duty Diesel Retrofits and Other Mobile Source Emission Reduction Projects and Programs," May 2007, and "Diesel Retrofit Technology: An Analysis of the Cost Effectiveness of Reducing Particulate Matter and Nitrogen Oxides Emissions from Heavy-Duty Nonroad Diesel Engines through Retrofits," May 2007, estimated a range of $18,700 to $87,600 per ton of PM emissions reduced, and a range of $1,900 to $19,000 per ton of NOx reduced for various retrofit scenarios.

and fuel technology strategies. The report also includes references to other sources of information on CMAQ program effectiveness. The report, including information on the source and availability of the methods is available online at: www.fhwa.dot.gov/environment/cmaqeat.

EPA has published a number of methodology guides for calculating emissions impacts of different types of strategies, notably diesel retrofits and program to reduce long-duration truck idling. The national Clean Diesel Program sponsored by EPA has published information and materials that relate to on- and off-road diesel engines. In particular, the Diesel Emissions Quantifier is an interactive tool developed by EPA to help State/local governments, fleet owners/operators, and others estimate emissions reductions and cost effectiveness for clean diesel projects. The Quantifier uses emissions factors and other information from EPA's National Mobile Inventory Model (NMIM) which includes the MOBILE 6.2 and NONROAD2005 models. For further information access: www.epa.gov/cleandiesel/publications.

A number of the methodologies identified through the review of the selected projects referenced these documents, particularly EPA guides and certification data related to emissions benefits of diesel retrofits and long-duration idle reduction.

STANDARDIZED APPROACHES FOR COMPARISONS OF PROJECTS

Some State DOTs and MPOs have developed their own guidebooks or emissions modeling tools to assist in documenting and evaluating proposed CMAQ projects and programs. These tools can help the State DOT or MPO in evaluating projects, reduce calculation errors, and ensure that local project sponsors provide information that is consistent and comparable with other similar projects. Several States have provided project sponsors with a spreadsheet into which sponsors can enter project-specific assumptions and receive back emissions benefits calculations. These guidebooks and tools often contain default parameters viewed as appropriate to the region.

Table 7 highlights several States and MPOs that provide emissions calculation aids or tools.

Table 7. Selected States and MPOs with Standardized Tools or Emission Calculation Methods.

State DOT or MPO	Standardized Tools or Emission Calculation Methods
Maricopa Association of Governments (Phoenix area, Arizona)	"Methodology for Evaluating Congestion Mitigation and Air Quality Improvement Projects" report provides standardized methodologies for calculating direct emissions effects (kg/day reduced) and cost-effectiveness at reducing emissions ($/metric ton)
Birmingham Regional Planning Commission (Alabama)	"A Guide for Estimating the Emissions Effects and Cost-Effectiveness of Projects Proposed for CMAQ Funding" includes standardized methodologies that are used to assess emissions impacts of different types of CMAQ projects, as well as cost-effectiveness
California	"Methods to Find the Cost-Effectiveness of Funding Air Quality Projects" guidebook and automated database contains standardized methods for estimating the emissions benefits and cost-effectiveness of different types of CMAQ projects; Access database files automate calculation procedures.
North Front Range MPO (Fort Collins area, Colorado)	A CMAQ Air Quality Benefit Program Excel workbook that includes a spreadsheet which allows project sponsors to select the Type, Area, and Category for the project being submitted. Based on those selections, the spreadsheet directs the sponsor to provide category-specific evaluation criteria and then it automatically

	calculates the emissions benefits and cost-effectiveness of the project. Two measures of cost-effectiveness are used: total current year project cost/annual emissions reduced, and CMAQ funds/annual emissions reduced. Although the calculation is automatic, within the workbook is another spreadsheet that provides the formulas used for the calculations.
Massachusetts Executive Office of Transportation	Excel workbook automatically calculates emissions benefits based on sponsor-provided assumptions; also provides sample air quality analysis methods.
New York State DOT	"CMAQtraq" application feeds into the DOT's database tool to determine air quality results. Local project sponsor provide input data with the application, and the DOT enters the information into the Microsoft Access database tool to determine the project's air quality results. The current version of CMAQtraq (ver. 6.2) has MOBILE6.2 emissions factors embedded in the calculations.
Pennsylvania DOT	"PAQONE" software analyzes a variety of transit, non-motorized travel, and roadway improvements using standardized methods
Wasatch Front Regional Council (Salt Lake City area, Utah)	Excel workbook automatically calculates emissions benefits based on default values or sponsor-provided assumptions.

Cost-Effectiveness Calculations

Analyzing the cost-effectiveness of CMAQ projects for both emissions reductions and congestion mitigation effects should be an important step in the project selection process, both in terms of the benefits that accrue to the States or MPOs receiving CMAQ funding and the net benefits achieved nationally by the funds distributed through the Federal CMAQ program. Broad statements about the types of projects that will or will not be funded in certain areas should be avoided because the types of strategies that are most cost-effective will vary due to local factors. Rather, States and MPOs may use cost-effectiveness calculations as a mechanism to objectively compare projects during review and selection. Examining project cost effectiveness can also be a way of bringing attention to the design or proposed application of the project, and can provide help in judging its suitability or most effective implementation strategy.

Several of the State DOTs and MPOs that provided project information for this study also had calculated cost-effectiveness, and were using standardize procedures to calculate cost-effectiveness. While the estimates of project duration, discount factors, and pollutants of concern varied, these methods allow projects to be evaluated across strategies and geographies to determine the most appropriate for funding.

For instance, in Alabama, standardized emissions calculation worksheets for common CMAQ strategies are provided to local project sponsors by the Birmingham MPO. The MPO, State DOT, and other agencies input information and assumptions for their projects into the spreadsheet model to determine travel impacts, emissions reductions, and cost effectiveness. The cost effectiveness is calculated using the following equation: Cost Effectiveness = (Annualized cost) / (Annual Emissions Reduction). Annualized costs include a 7 percent discount rate and a capital recovery factor to account for the project service life multiplied by the total capital cost of the project to estimate the average annual cost. Cost effectiveness calculations are provided for HC, NOx, PM_{10}, (HC + NOx), (PM_{10} − NOx), and (PM_{10} + NOx) in both dollars per lb per year and dollars per kg per year. In the case of the North Front Range MPO in Colorado, two measures of cost-effectiveness were calculated: total project cost/annual emissions reduced, and CMAQ funds/annual emissions reduced. The second calculation takes into account the source of project funding, and enables projects with a higher non-CMAQ funding share to shower better cost-effectiveness.

Transparent, Inclusive Selection Processes

While the CMAQ program is intended to enable local agencies the flexibility to select projects that meet the transportation infrastructure, political, and geographic needs of local areas, FHWA CMAQ Interim Guidance includes language requiring that,

> "The CMAQ project selection process should be transparent, in writing, and publicly available. The process should identify the agencies involved in rating proposed projects, clarify how projects are rated, and name the committee or group responsible for making the final recommendation to the MPO board or other approving body."[50]

Although the collection of data did not reveal readily available documentation of the CMAQ project selection process in most cases, it did identify several States and MPOs that appear to have consistent and robust project selection procedures. In addition, SAFETEA–LU encourages State DOTs and MPOs to consult with State and local air quality agencies about the estimated emissions reductions from CMAQ proposals. States which seek guidance and/or evaluation assistance from these agencies will also ensure more accurate air quality analyses for CMAQ projects. Table 8 provides examples of State DOTs and MPOs that appear to have documented, transparent project selection methods.

Table 8. Identified States and MPOs with Transparent Project Selection Methods.

State DOT or MPO	Selection Process
North Front Range MPO (Fort Collins, Colorado)	Current process utilizes a three-tiered scoring system to rank projects: 50 percent of the score is assigned to short-term air quality impacts (rankings based on VMT and carbon monoxide reduction estimates for year one); 20 percent for long-term benefits (estimated for years two through five), and 30 percent for bonus features (e.g., overmatch, multi-agency or public/private partnerships, and multi-modal projects). Standardized calculation procedures are used to analyze emissions effects. CMAQ Project Selection Committee includes representatives form the Colorado Department of Transportation, Colorado Air Pollution Control Division, FHWA, FTA, and U.S. EPA. Selection Committee develops list of recommended projects based on project scoring, as well as other intangible elements (which may include regional equity, project readiness, synergies with projects funded from STP or other sources, and project mix).
Hillsborough MPO (Tampa, Florida)	CMAQ projects are evaluated by a committee of representatives of the MPO, FDOT, Florida Department of Environmental Protection, and the Hillsborough County Environmental Protection Commission (EPC) based on a series of qualitative and quantitative measures. Final project ranking is based on the average total score

---

[50] Sec. 1808: Addition to CMAQ Eligible Projects. Publication of Interim Guidance on the Congestion Mitigation and Air Quality Improvement (CMAQ) Program. Dec. 19, 2006 Federal Register.

	assigned by each of the four reviewing agencies. The ranking is based on 5 criteria, each scored on a scale one to five: 1) projects that remove vehicles from the road or reduce travel delay; 2) outreach projects that change the public's driving behaviors; 3) projects with the most efficient dollar per ton cost/benefit figure for reducing NOx; 4) projects with air quality benefits to be realized within 3 years of funding; and 5) projects identified in CMS Study and/or 2025 LRTP Interim Plan.
Georgia DOT	A project selection process, developed by GDOT, the Environmental Protection Division, the Georgia Regional Transportation Authority, and Georgia Environmental Facilities Authority – together known as State Air Quality Partners – is used consistently across the State. Previously, CMAQ funds were confined to Atlanta area, but with PM2.5 designations, a new approach was developed. The process does not sub-allocate funding to specific MPOs but instead seeks to support the most beneficial projects for reducing emissions and meeting air quality goals across the state.
Rouge Valley MPO (Oregon)	The Rouge Valley MPO awards points for meeting certain evaluation criteria outlined in a Project Evaluation Questions & Intent questionnaire form. Criteria include emissions reduction, and other considerations, such as: long-term air quality improvement, potential to reduce reliance on automobiles, potential to mitigate congestion, completes a multi-modal transportation system, located in city limits or inside Urban Containment Boundary, and diesel retrofits. Points awarded for the criteria are used to develop an overall score for each project.
Southwestern Pennsylvania Commission (Pittsburgh MPO)	The MPO provides potential project sponsors with a CMAQ application and instruction package, including schedule, guidelines, and selection criteria to guide sponsors of candidate projects through the CMAQ process. Application forms can be filled out electronically. Candidate CMAQ projects are placed into appropriate investment categories and sent to appropriate SPC members, PennDOT Districts, and transit agencies as well as SPC's CMAQ Evaluation Committee. Projects are evaluated for effects on emission and cost-effectiveness based on standardized models developed for PennDOT. A scorecard is completed by SPC staff for each project rating each candidate project on consistency with priority project types (e.g., diesel retrofits, traffic signal improvements, TDM, commuter bicycle/pedestrian improvements) and 9 ancillary selection factors to develop total weighted score. CMAQ Evaluation Committee members use this information to develop recommendations for each investment category.
Wasatch Front Regional Council (Salt Lake City, Utah MPO)	The MPO has adjusted its evaluation criteria and procedures over time. In the past, a score was calculated using a weighted ranking system that considered the following: (10%) Project in a congested corridor (15%) Length (years) of project effectiveness (25%) Emissions reduction (25%) Congestion reduction (VHT) (25%) Cost This objective ranking was then combined with subjective rankings by staff and 3 different committees consisting of city planners and elected officials. Currently, the MPO uses air quality cost-effectiveness rankings as a primary criterion for project selection within different categories of projects. The MPO generally allocates a certain percentage of funding for each major project category (e.g.,

	bicycle/pedestrian projects, transit projects) in order to ensure a variety of project types are implemented, and ranks project cost-effectiveness within each category. Field visits are also conducted of projects proposed for funding.

Collecting Project-Specific Data and Conducting Project Evaluation Studies

Regardless of the model or methodology used to calculate emissions benefits of CMAQ-funded projects, good inputs are needed to produce good outputs. States and MPOs should take efforts to gather data through surveys and other data collection methods to justify and/or make assumptions. Some States, such as Michigan and New York, require project sponsors to provide the source and justification of all inputs and assumption used in the emissions calculations. Not only does this ensure that the project demonstrates an air quality benefit, but it allows the State to evaluate the accuracy of the analyses. Many of the project samples cited local data in their calculations, including factors such as average trip lengths, park-and-ride utilization rates, number of actual vanpool riders, transit riders, number of rideshare matches, etc.

A comparison of forecasted impacts (via project selection methodologies) to actual results (based on ex post evaluation) can help inform the rigor and accuracy of calculation methodologies and project selection procedures. State DOTs report the status and effectiveness of the CMAQ programs in their States to the U.S. Department of Transportation. The information from these reports is entered into the CMAQ database and can provide States with an effective tool for monitoring and evaluating the results of CMAQ-funded projects. Performing project evaluation studies allows States to periodically review their project selection criteria to ensure it remains appropriate and up-to-date. Evaluation studies also provide States and MPOs with new and more accurate data to be used in future emissions analysis calculations. For example, data on the actual speed improvements along freeways due to an ITS system implementation may lead to an increase in the baseline speeds along freeways in the entire region.

Although post-project analysis is not commonly conducted on a program-wide basis for all CMAQ projects within a state or MPO area, in some cases, post-project evaluations are conducted by States and MPOs for specific projects or types of projects, especially those that are included as part of a regional conformity analysis. Post-project evaluation is a good practice in helping to provide information on the accuracy of emissions forecasts and assumptions used in emissions calculations. In some cases, however, post-project analysis may not be practical, such as for a small project where conducting a rigorous evaluation might cost nearly as much as the project itself. Some examples of post-project evaluations that have been conducted are listed in Table 9 below.

Table 9. Selected States and MPOs that Conduct or Call for Post-Project Analysis.

State or MPO	Types of Post-Project Analysis Conducted
California	The California Air Resources Board uses post-project evaluation reports generated by regional air districts as part of a state grant process to update the California emissions methodology guidebook.
Georgia DOT	Detailed evaluations have been conducted for the regional Clean Air/TDM program.
New York State DOT	New York State DOT conducts an annual evaluation of its Clean Air / Ozone Action Days outreach program.

Metropolitan Washington Council of Governments (Washington, DC area)	MWCOG conducts a regular evaluation of its Transportation Emissions Reduction Measures (TERMs), which include a number of programs funded in part by CMAQ (from allocations from Maryland, Virginia, and the District of Columbia). The TERMs report includes collection of data on participation rates in programs, including collection of survey data.

Next Steps

The study team's collection of data on the selected set of projects revealed a number of strengths and limitations in the analysis of CMAQ projects. On the one hand, many of the project analyses were conducted based on relatively limited data, using sketch planning methodologies, with limited verification of results. This is perhaps not surprising given the limited scope of many projects, limited data and tools available for analyzing many of these projects, and the costs and effort associated with conducting detailed evaluation studies. On the other hand, it appears that a number of states and MPOs have implemented good practices to help standardize the emissions analyses, collect local data for use in calculations, rank project cost-effectiveness, and implement systematic procedures for evaluation. These procedures often take into account multiple factors beyond emissions reduction cost-effectiveness.

In Phase II of this evaluation project, FHWA, in consultation with EPA, conducted a set of limited on-site case studies and/or program analyses. These case studies add to the national understanding of how the CMAQ program operates at the state and local levels, and may build on five case studies (Los Angeles, Chicago, Houston, Washington, DC, and Albany) conducted as part of the TRB study on the CMAQ Program. The Phase II case studies provide information on how States and MPOs are analyzing, prioritizing, and selecting projects, and implementing the CMAQ program to meet State and local objectives. The insights gained from these studies will help to inform States and MPOs about best practices, and a variety of potential ways to improve the effectiveness of their CMAQ program implementation efforts.

APPENDIX A. LIST OF STATE AND LOCAL CONTACTS

STATE	CONTACT	AGENCY	EMAIL
Alabama	Harry Hee	Birmingham DOT	hhe@rpcgb.org
Arizona	Beverly Chenausky	Arizona DOT	bchenausky@azdot.gov
Arizona	Dean Giles	Maricopa Association of Governments	dgiles@MAG.maricopa.gov
California	Jody Tian	Caltrans	Jody.Tian@dot.ca.gov
California	Sookyung Kim	SANDAG	ski@sandag.org
California	Raquel Pacheco	KCOG	rpacheco@kerncog.gov
California	Harold Brazil	MTC	hbrazil@mtc.ca.gov
California	Jason Paukovits	COFCG	jasonp@fresnocog.org
Colorado	Steve Cook	Denver Regional COG	scook@drcog.org
Colorado	Tia Raamot	North Front Range MPO	tramot@nfrmpo.org
Connecticut	Judy Raymond	Connecticut DOT	Judy.Raymond@po.state.ct.us
District of Columbia	Daivamani Sivasailam	Metropolitan Washington COG	siva@mwcog.org
Florida	David Lee	Florida DOT	David.Lee@dot.state.fl.us
Florida	Brian Pessaro	Florida DOT	Brian.Pessaro@dot.state.fl.us
Idaho	Lisa Josleyn	Idaho DOT	Lisa.Josleyn@itd.idaho.gov
Indiana	Cory Hull	Mid-America COG	chull@macog.com
Indiana	Jerry Halperin	Indiana DOT	JHalperin@indot.IN.gov
Indiana	Gary Evers	Northwestern Indiana RPC	gevers@nirpc.org
Kentucky	David Schaars	Lexington Area MPO	Davids3@lfucg.com
Louisiana	Huey Dugas	Baton Rouge MPO	hdugas@brgov.com
Maine	Duane Scott	Maine DOT	Duane.scott@maine.gov
Maryland	Howard Simons	Maryland DOT	hsimons@mdot.state.md.us
Massachusetts	Ethan Britland	Executive Office of Transportation	Ethan.Britland@eot.state.ma.us
Michigan	Pete Porciello	Michigan DOT	Porciellop@michigan.gov
Missouri	James Joerke	Mid-America RC	Jjoerke@marc.org
New York	John Zamurs	New York DOT	jzamurs@dot.state.ny.us
New York	Christa Ippoliti	New York DOT	cippoliti@dot.state.ny.us
Ohio	Dave Moore	Ohio DOT	Dave.Moore1@dot.state.oh.us
Ohio	Andy Reser	Ohio Kentucky Indiana Regional COG	areser@oki.org
Oregon	Matt Hermen	Rogue Valley COG	mhermen@rvcog.org
Pennsylvania	Michael Baker	Pennsylvania DOT	michaelba@state.pa.us
Rhode Island	Katherine Trapani	State Planning Council	
Tennessee	Mike Conger	Knoxville Regional TPO	Mike.Conger@knoxtrans.org
Tennessee	Matt Meservy	Nashville MPO	Matt.meservy@nashville.gov
Texas	Christie Jestis	North Central Texas COG	cjestis@nctcog.org

Texas	Shelley Whitworth	Houston-Galveston Area Council	Shelley.whitworth@h-gac.com
Texas	Andrew DeCandis	Houston-Galveston Area Council	Andrew.DeCandis@h-gac.com
Utah	Kip Billings	Wasatch Front RC	kbillings@wfrc.org
Washington	Kelly McGourty	Puget Sound RC	kmcgourty@psrc.org
Wisconsin	John Duffe	Wisconsin DOT	John.Duffe@dot.state.wi.us

APPENDIX B. EMISSIONS FACTORS AND ASSUMPTIONS USED IN EMISSIONS REDUCTION COST-EFFECTIVENESS CALCULATIONS

Most CMAQ projects and programs can be analyzed in multiple ways, and variations of these approaches are available. The methods described in this report are generally simple sketch planning approaches that involve three main factors: 1) estimating the travel, speed, or vehicle changes associated with the strategy; 2) estimating emissions impacts; and 3) calculating cost effectiveness. The cost-effectiveness calculation is described in further detail in Section 4. This appendix describes the emissions factors used to recalculate normalized emissions reductions using data provided by the project sponsor.

Unless otherwise noted in each of the report sections, all of the on-road projects presented in this report were recalculated using emissions factors generated from MOBILE6.2. Factors were generated using standard defaults for Year 2008. Emissions were generated for start (trip-based factors assuming 100 percent cold start) and running emissions (per mile factors). The recognition of a difference between trip starts emissions and running emissions is significant, since emissions control equipment does not function as effectively from a "cold start" causing the release of more pollutants during the first few miles of a trip. Finally, the modeling employed the "NO REFUELING" command in MOBILE6.2, since refueling emissions are associated with gas stations and are not normally affected by the types of projects outlined in this report.

Guidance is available from EPA and FHWA on the use of MOBILE6.2 for further information.

Table B-1. Major Input Parameters for MOBILE6.2 Emissions Factor Modeling.

Parameter or Variable	Values or Sources
Vehicle Fleet and Activity Inputs	
VMT mix	EPA national average (default)
Mileage accrual rates	EPA national average (default)
Vehicle model year (registration) distribution	EPA national average (default)
Diesel sales fractions	EPA national average (default)
Soak time distribution	EPA national average (default), or All soak times >720 minutes (corresponds to 100% cold starts).
Starts per day distribution	EPA national average (default), or Zero starts per day (for running emissions only)
Region	Low altitude
Vehicle speeds	Varied 2.5 mph and 3-65 mph by integers, with single average speed per scenario.
Roadway facility (functional classes)	Arterial (allows use of specific average speeds)
Seasonal/Meteorological Inputs	
Month of evaluation	July
Temperatures for all pollutants	Minimum: 68.0° F Maximum: 94.0° F (Representative summer temperatures only. Actual source for these values is high-ozone-day data from Boston, MA nonattainment area SIP.)
Absolute humidity	MOBILE6.2 default
Fuel Inputs	

Parameter or Variable	Values or Sources
ASTM Class	MOBILE6.2 default
Oxygenated fuels	No (MOBILE6.2 default)
Reformulated gasoline	No (MOBILE6.2 default)
Gasoline RVP	8.7 psi (Representative summer RVP only. Actual source for this value is Philadelphia, PA nonattainment area SIP.)
Diesel fuel sulfur content	15 ppm
State Program Inputs	
Inspection/Maintenance (I/M) Program	No program (MOBILE6.2 default)
Low Emitting Vehicle (LEV) Program	No program (MOBILE6.2 default)
Anti-tampering program (ATP)	No program (MOBILE6.2 default)
Stage II refueling controls	Not modeled (NO REFUELING command used).
Other Inputs	
Particulate matter emissions parameters	EPA national average (default)
All other inputs	EPA national average (default)

APPENDIX C. CMAQ PROJECT TEMPLATES

This appendix provides information about the reviewed projects gathered in the research phase of this study. The information provided by project sponsors was transcribed into individual project "templates." The project template was designed to compile all the critical detail about particular project facts in one place to ease subsequent reviews, comparisons, and analysis.

The project templates record the following information reported by project sponsors about each CMAQ project:

- Indentifying Information: Category and Subcategory, State, Year, and CMAQ ID number;

- Description of critical project characteristics and background;

- Impacts on travel: change in vehicle trips, vehicle miles traveled (VMT), transit trips, creation of vanpools, and congestion (speed and delay);

- Emissions reductions: change in emissions of volatile organic compounds (VOC), NOx, CO, $PM_{2.5}$, and PM_{10}, measured in kilograms per day; and

- Project Costs and cost-effectiveness: Capital (annualized) and operating costs, from CMAQ and non-CMAQ sources, Annualized costs, and a cost-effectiveness calculation, if provided by the local project sponsor.

The profiles were designed to record supporting information concerning the methodologies employed in any of the steps (travel, emissions, costs), such as assumptions, time frames, service lives, and discount rates. Notes were also entered at the bottom of the templates to document the general quality of the information and to note any discrepancies between the CMAQ database and the information provided by the project sponsor.

Category: **TRAFFIC FLOW IMPROVEMENTS**		Subcategory: **Traffic Signalization**
CMAQ Project ID: MI20020058		Project Year: 2002
Location: Macomb County, Michigan		MPO: Southeast Michigan Council of Governments

Description: **Signal Timing along Ryan Rd. 8 Mile to 23 Mile -** This project will fund the coordination of traffic signals along Ryan Rd. from 8 mile to 23 mile in Warren, Sterling Heights, and Shelby Township in Michigan. Macomb County borders the City of Detroit to the South and Lake St. Clair to the east; Ryan Rd. serves as a major North-South arterial in the area. As a result of this project, vehicle travel speeds are expected to increase 4 mph during both peak and off-peak periods.

TRAVEL IMPACTS

		METHODOLOGY/ASSUMPTIONS:
Δ Vehicle trips:	NA	Miles of urban minor arterial affected: 15 miles
Δ VMT:	NA	Daily, 2-way traffic volume = 23,519 vehicles with 40% of travel occurring in peak periods.
Δ Speed:	+ 4 mph	Peak VMT = 15 miles * 23,519 vehicles * 0.4 = 141,114 miles
Δ Delay:	NA	Off Peak VMT = 15 miles * 23,519 vehicles * 0.6 = 211,671 miles
Δ SOV	NA	
Δ CP/VP	NA	
Δ Transit	NA	Travel Speeds before project are 31 mph in peak, and 41 mph in off-peak.
Δ Walk	NA	Travel Speeds after project are 35 mph in peak, and 45 mph in off-peak.
Δ Bike	NA	

EMISSIONS

		METHODOLOGY/ASSUMPTIONS:
Δ VOC	- 40.076 kg/day	Emissions reductions calculated using Mobile 5a running emissions factors (g/mile) for VOC at the following speeds:
Δ NO$_X$	NA	Peak: 31 mph: VOC = 1.843 35 mph: VOC = 1.697
Δ CO	NA	Off Peak: 41 mph: VOC = 1.526 45 mph: VOC = 1.434
Δ PM$_{10}$	NA	
Δ PM$_{2.5}$	NA	
Δ Total	- 40.076 kg/day (0.0442 tpd)	Calculate daily emissions reduced = (change in peak emissions * Peak VMT) + (change in off-peak emissions * Off-peak VMT)
		VOC Emissions = ((1.697 - 1.843) * 141,114 miles) + ((1.434 - 1.526) * 211,671 miles) / 1,000 = - 40.076 kg/day

COSTS

				Project life:__10__ yrs	Interest rate:___7__%

	CMAQ	NON-CMAQ	TOTAL	METHODOLOGY/ASSUMPTIONS:
Capital	$660,000	$0	$660,000	Materials provided by the local sponsor indicate no local match for this project.
Adm/oper	$0	$0	$0	
Total	$660,000	$0	$660,000	
Total annualized public cost:	$110,256			The cost effectiveness analysis provided by the project sponsor assumes the service life of the project and amortization period are 15 years.
Annual revenues:	None			
Net public cost:	$660,000			
Annual private cost:	NA			
Total net cost	$660,000			

NOTE: Emissions reductions reported by the project sponsor do not match the emissions reductions in the CMAQ database (-57 kg/day VOC).

Category: **TRAFFIC FLOW IMPROVEMENTS**		Subcategory: **Traffic Signalization**
CMAQ Project ID: LA20040001		Project Year: 2004
Location: Baton Rouge, Louisiana		MPO: Capital Regional Planning Commission

Description: **Continuous Flow Intersection at Airline and Sherwood Forest Blvd.** - The project involves the modification of an intersection – Airline Highway @ Sherwood Forest Blvd. – in order to increase traffic flow and reduce congestion and delay using an innovative intersection improvement concept called continuous flow intersection (CFI). This concept eliminates volume build-up due to the left-turn cycle of the traffic signals by moving the left-turn out of the main intersection, thus allowing through-traffic and left-turning traffic to move through the intersection at the same time. The improvements will reduce total traffic delay by 3 hours during both the morning and evening peak hours. The improvements will also enhance traffic flow and reduce emissions during off-peak times, but the benefit will be greatest during peak hours.

TRAVEL IMPACTS

		METHODOLOGY/ASSUMPTIONS:
ΔVehicle trips:	NA	Travel analysis performed by VISSIM Microscopic Simulation model used recent traffic counts and traffic signal information to give an average delay in seconds per vehicle at the intersection. (Total Delay = Peak Hour volume * Average Delay in sec/veh/3600.) Hourly reduction in delay is calculated separately for the AM and PM peak periods and summed.
ΔVMT:	NA	
ΔSpeed:	NA	
ΔDelay:	- 388 vehicle-hours/hour	
ΔSOV	NA	The analysis showed that the proposed improvements would enhance traffic flow during peak hours – increasing from 5,800 to 6,500 VPH in the AM peak, and from 6,200 to 6,700 VPH in the PM peak. Average delay would drop from 92.6 to 36.0 sec/veh in the AM peak, and 178.3 to 34.4 sec/veh in the PM peak. Net reduction in delay is 84.2 veh-hr/hr in the AM peak and 304.5 veh-hr/hr in the PM peak. Intersection analysis also shows change in LOS for each intersection segment.
ΔCP/VP	NA	
ΔTransit	NA	
ΔWalk	NA	
ΔBike	NA	

Emissions

		METHODOLOGY/ASSUMPTIONS:
Δ VOC	- 20.12 kg/day	Emissions reductions calculated from changes in delay.
Δ NOx	- 5.18 kg/day	Emissions factors were developed using MOBILE6, using 2.5 Mph speed, and converted into idle emissions factors.
Δ CO	NA	Emissions factor for VOC = 10.35 g/mi
Δ PM10	NA	Emissions factor for NOX= 2.67 g/mi
Δ PM2.5	NA	
ΔTotal	- 25.30 kg/day (0.028 tpd)	Emissions reduction = Delay in vehicle-hours/hour * Emissions Factor * 2.5 (conversion of gm/mi to gm/hr) * 2 hours per day (calculated for 2-hour Am peak and 2-hour Pm peak separately, and summed)

Costs

				Project life:__ 10 _ yrs	Interest rate: __ 7 __%
	CMAQ	Non-CMAQ	Total	Methodology/Assumptions:	
Capital	$4,400,000	$1,100,000	$5,500,000		
Adm/oper	$0	$0	$0	Assumes that project has benefits 260 days per year. To calculate overall cost-effectiveness, need to develop assumptions regarding useful life of project (could be 20 years for an infrastructure project of this nature, although it's not clear that delay reductions will remain constant over this long of a period).	
Total	$4,400,000	$1,100,000	$5,500,000		
Total annualized public cost:			$904,773		
Annual revenues:			NONE		
Net public cost:			$5.5 M		
Annual private cost			NA		
Total net cost			$5.5 M		

Category: **TRAFFIC FLOW IMPROVEMENTS**				Subcategory: **Traffic Signalization**
CMAQ Project ID: KY20050008			Project Year: 2005	
Location: Lexington, Kentucky			MPO: Lexington Area MPO	

Description: **Fiber Optic Cable Installation For Traffic Signal Optimization** - This project will fund an expansion of the fiber optic cable installation for the arterial road network in Fayette County. Fiber optic cables provide a more reliable and dependable communication medium for the current traffic signal system data and video needs. It also provides the needed communications infrastructure into the foreseeable future for roadside subsystems like vehicle detection and surveillance. Fiber optic cable installation greatly reduces maintenance requirements of the existing, aging copper wire and analog telephone communication infrastructure and it all but eliminates the interruption of service due to lightning strikes and electrical power surges. Thus, this technology has helped to eliminate most of the need to block traffic lanes for repairs, thereby improving the safety of the roadway for all users and lessening delays caused by lane blockages.

TRAVEL IMPACTS

		METHODOLOGY/ASSUMPTIONS:
Δ Vehicle trips:	NA	Delay = 4 minutes per vehicle, based on a report on the Integrated Traffic Signal System from 2001, which determined the reduction in delay and corresponding emissions savings by using an average reduction for 18 intersections and projecting it throughout the total system.
Δ VMT:	NA	
Δ Speed:	NA	
Δ Delay:	- 4 min/vehicle	
Δ SOV	NA	Vehicle counts provided by Kentucky Transportation Cabinet, Division of Planning.
Δ CP/VP	NA	
Δ Transit	NA	
Δ Walk	NA	
Δ Bike	NA	

EMISSIONS

		METHODOLOGY/ASSUMPTIONS:
Δ VOC	- 35.5 kg/day	Emissions factors are based on EPA calculations for general vehicle fleet mix. The percentage of vehicle types or classifications was used to determine the grams of pollutant reduced per minute by the reduction in delay, using Our Nation's Highways, from the Federal Highway Administration (FHWA).
Δ NOx	- 9.1 kg/day	
Δ CO	- 378.0 kg/day	
Δ PM10	NA	
Δ PM2.5	NA	
Δ Total	- 44.6 kg/day (0.05 tpd)	(average of vehicle counts per arterial) x (minute of delay reduced by fiber optic) x (g/min per VOC, NOx, CO) = total grams VOC, NOx, CO per day

COSTS

				Project life:__ NA _ yrs	Interest rate: __ 7 __%
	CMAQ	NON-CMAQ	TOTAL	METHODOLOGY/ASSUMPTIONS:	
Capital	$320,000	$80,000	$400,000	Cost-effectiveness was not provided by the project sponsor.	
Adm/oper	$0	$0	$0		
Total	$320,000	$80,000	$400,000		
Total annualized public cost:	NA				
Annual revenues:	None				
Net public cost:	$400,000				
Annual private cost	NA				
Total net cost	$400,000				

NOTE: Emissions reductions reported in CMAQ database differ from estimates provided or calculated from sponsor-provided documentation. Reductions reported in the CMAQ database were reported in the template. The project calculation showed much higher values (-200.94 kg/day VOC, -54.89 kg/day NOx, -2,272 kg/day CO); however, a more recent, similar project reported figures closer to the values reported in the CMAQ database. The project specifics seem to indicate that the delay reduction (4 min/vehicle) is an extrapolation of the effects of the project across the entire system, not just for the 18 intersections.

Category: **TRAFFIC FLOW IMPROVEMENTS**	Subcategory: **Traffic Signalization**

CMAQ Project ID: OH20050033	Project Year: 2005
Location: Newark, Ohio	MPO: Licking County Area Transportation Study

Description: **Signal Timing along West Main Street** – This project will coordinate the signals at intersections along Main Street in Newark, Ohio. The data for this study were taken directly from a 1999 traffic study, which determined which intersections should qualify for signal timing to reduce the amount of delay and emissions. The intersections are Main street and Williams Street, Fulton Street, SR79 south bound ramps, SR79 north bound ramps, Union Street, and Eleventh Street.

TRAVEL IMPACTS

		METHODOLOGY/ASSUMPTIONS:
Δ Vehicle trips:	NA	Modeling done using Syncro Version 5 for the PM period. There is minor congestion during the entire business day, so the AM and PM Peak Periods are similar. The change in delay was calculated for four (4) intersections using the following formula: (Approach Volume * Stop Control Delay) – (Approach Volume * Signalized Delay) = Total Delay Reduction
Δ VMT:	NA	
Δ Speed:	NA	
Δ Delay:	- 702 hours/day	
Δ SOV	NA	SR79S (1,568 veh * 26.1 veh/sec) – (1,568 veh * 6.5 veh/sec) = -8.54 hours/day
Δ CP/VP	NA	SR79N (1,523 veh * 1,579 veh/sec) – (1,523 veh * 15.5 veh/sec) = -661.7 hour/day
Δ Transit	NA	Union St. (1,580 veh * 9.2 veh/sec) – (1,580 veh * 22.6 veh/sec) = +5.88 hours/day
Δ Walk	NA	11th St. (1,536 veh * 95.4 veh/sec) – (1,568 veh * 7.2 veh/sec) = -37.63 hour/day
Δ Bike	NA	

EMISSIONS

		METHODOLOGY/ASSUMPTIONS:
Δ VOC	- 5.115 kg/day	Emissions reductions calculated using Mobile6. Idle emissions calculated using exhaust emissions for a 2.5 mile/hour average speed.
Δ NOx	- 3.909 kg/day	
Δ CO	- 90.710 kg/day	
Δ PM10	NA	The Mobile Factors used Main Street as a Minor Arterial Urban – Class 16 and all intersecting streets as Local Urban – Class 19 to determine emissions.
Δ PM2.5	NA	
Δ Total	9.02 kg/day (0.0099 tpd)	

COSTS

				Project life: _5-10_ yrs	Interest rate: _7_%

	CMAQ	NON-CMAQ	TOTAL	METHODOLOGY/ASSUMPTIONS:
Capital	$355,302	$284,241	$639,543	Documentation provided by the State indicates the project was funded in FY 2005 and 2006 and some portion of funds de-obligated in FY 2007. Cost-effectiveness was not provided by sponsor.
Adm/oper	$0	$0	$0	
Total	$355,302	$284,241	$639,543	
Total annualized public cost:	$98,414			

			FFY 05	FFY 06	FFY 07	Total Obligated
Annual revenues:	None		$55,214.50	$359,563.00	-$59,475.57	$355,301.93
Net public cost:	$639,543					
Annual private cost	NA					
Total net cost	$639,543		Project sponsor assumes service life is 5 years. The cost-effectiveness analysis in this study used 10 years for consistency with other signalization projects.			

Category: **TRAFFIC FLOW IMPROVEMENTS**		Subcategory: **Traffic Signalization**
CMAQ Project ID: TN20050016		Project Year: 2005
Location: Knoxville, Tennessee		MPO: Knoxville Urbanized Area MPO
Description: **Signal Timing on SR 169 Cedar Bluff to College St.** - This project will fund the traffic signal timing and synchronization of traffic signals along Middlebrook Park from Cedar Bluff St. to College St.		

TRAVEL IMPACTS		
Δ Vehicle trips:	NA	**METHODOLOGY/ASSUMPTIONS:**
Δ VMT:	NA	Daily VMT = 25,935 average daily traffic x 9.47 mile corridor length = 245,065 VMT on corridor.
Δ Speed:	+ 4 mph	
Δ Delay:	NA	An average improvement in speed/travel of 12% for traffic signal upgrades of this type is noted in the publication "A Toolbox for Alleviating Traffic Congestion and Enhancing Mobility" from ITE.
Δ SOV	NA	
Δ CP/VP	NA	
Δ Transit	NA	Average speed increased from 34 mph to 38 mph.
Δ Walk	NA	
Δ Bike	NA	

EMISSIONS		
Δ VOC	- 14.969 kg/day	**METHODOLOGY/ASSUMPTIONS:**
Δ NOₓ	+ 2.206 kg/day	Emissions factors for before project implementation and after project implementation based on MOBILE6 and average speeds of 34 mph and 38 mph, respectively.
Δ CO	NA	
Δ PM₁₀	NA	
Δ PM₂.₅	NA	
Δ Total	- 12.763 kg/day (0.014 tpd)	Emissions reduction = VMT x (Emissions Factor before project – Emissions Factor after project)
		VOC Emissions reduction = 245,065 VMT x (1.883 – 1.826) / 1000 = 14.969 kg/day
		NOx Emissions reduction = 245,065 x (1.847 – 1.856) / 1000 = -2.206 kg/day

COSTS			
		Project life:__10__ yrs	Interest rate: ___7__ %

	CMAQ	NON-CMAQ	TOTAL	METHODOLOGY/ASSUMPTIONS:
Capital	$33,000	$0	$33,000	Cost-effectiveness was not provided by the project sponsor.
Adm/oper	$0	$0	$0	
Total	$33,000	$0	$33,000	
Total annualized public cost:	$5,078			
Annual revenues:	None			
Net public cost:	$33,000			
Annual private cost	NA			
Total net cost	$33,000			

Strategy: **TRAFFIC FLOW IMPROVEMENTS**		Category: **Traffic Signalization**
CMAQ Project ID: KY20060009		Project Year: 2006
Location: Kentucky		MPO: Lexington Area MPO

Description: Installation of Reversible Lanes on Nicholasville Road (US 27) - This project will create a third northbound traffic lane for the morning peak period using reversible lane controls on Nicholasville Road (US 27) from Southpoint Drive to Tiverton Way. By taking advantage of unutilized median space and low early morning left-turning volumes at the intersection, reversible lane control methods can be used to reassign one of the left-turn lanes as a through-lane during the high-volume period. The project will also require the expansion of the computerized traffic signal system to add new reversible lane signals. This project will improve the traffic flow on Nicholasville Road, which will in turn reduce traffic congestion, accidents, and delays, and ultimately improve air quality.

TRAVEL IMPACTS

		METHODOLOGY/ASSUMPTIONS:
Δ Vehicle trips:	NA	Delay (vehicle-hours):
Δ VMT:	NA	2006 No-build = 362 vehicle-hours of delay
Δ Speed:	NA	2006 Build = 299 vehicle-hours of delay
Δ Delay:	- 63 vehicle-hours	Change in delay due to project implementation = 362 - 299 = 63 vehicle-hours = 17% reduction in delay.
Δ SOV	NA	
Δ CP/VP	NA	
Δ Transit	NA	Reduction in delay determined by the Synchro model output, based on a one-hour simulation. These one hour peak delay reductions, per day, were used to determine an average delay for two hours of peak travel reductions.
Δ Walk	NA	
Δ Bike	NA	

EMISSIONS

		METHODOLOGY/ASSUMPTIONS:
Δ VOC	- 2.889 kg/day	The delay reductions were used to calculate the emissions savings using emissions factors provided by US EPA Office of Transportation and Air Quality.
Δ NO$_x$	- 1.089 kg/day	
Δ CO	- 11.95 kg/day	
Δ PM$_{10}$	NA	Reduction in delay * average of vehicle mix for kg/min per CO, NOx, VOC * 255 days per year.
Δ PM$_{2.5}$	NA	
Δ Total	- 4 kg/day (0.0044tpd)	

COSTS

Annualized public costs				Project life:__ 10 _ yrs	Interest rate: __7__%
	CMAQ	NON-CMAQ	TOTAL	METHODOLOGY/ASSUMPTIONS: Assumes benefits 255 days/year.	
Capital	$400,000	$100,000	$500,000		
Adm/oper	$0	$0	$0	Cost-effectiveness was not provided by project sponsor.	
Total	$400,000	$100,000	$500,000		
Total annualized public cost:	$74,536				
Annual revenues:	None				
Net public cost:	$500,000				
Annual private cost	NA				
Total net cost	$500,000				

Category: **TRAFFIC FLOW IMPROVEMENTS**	Subcategory: **Traffic Signalization**

CMAQ Project ID: Not Yet Assigned	Project Year: 2007
Location: Albany, New York	MPO: Capital District Transportation Committee (CDTC)

Description: Construction of a Two Lane Roundabout at Fuller and Washington - This project will fund the construction of a two-lane roundabout at the intersection of Washington Avenue and Fuller Road (County Road 156) in the City of Albany, Albany County. The intersection currently operates under the control of a traffic signal. The roundabout intersection will include the construction of sidewalks.

TRAVEL IMPACTS

Δ Vehicle trips:	NA
Δ VMT:	NA
Δ Speed:	+ 14 mph
Δ Delay:	- 6.5 sec/veh.
Δ SOV	NA
Δ CP/VP	NA
Δ Transit	NA
Δ Walk	NA
Δ Bike	NA

METHODOLOGY/ASSUMPTIONS:

48,670 average traffic volumes for Year 2009 were calculated using the CDTC STEP Model. The CDTC STEP Model forecast was validated using a 1999 intersection count and used to calculate seconds of delay for approach vehicles with the existing signalized intersection. The RODEL Roundabout Capacity Model was used to conduct an analysis of the Washington Avenue/Fuller Road intersection and was used to calculate seconds of delay for approach vehicles under the new, roundabout build scenario. (11.5 sec avg "No Build" delay – 5 sec avg "New Roundabout" delay = 6.5 sec avg change in delay.

Washington Ave and Fuller Rd Roundabout Capacity RODEL Analysis   1999 counts

Leg	Flow (veh/hr)	Avg Delay	Avg Queue	Max Delay	Max Queue
Washington Ave EB	1212	5 sec	2	7 sec	2
Fuller Rd NB	591	5 sec	1	7 sec	1
Washington Ave WB	1368	6 sec	2	9 sec	3
Fuller Rd SB	885	4 sec	1	6 sec	1

Washington Ave and Fuller Rd Roundabout Capacity RODEL Analysis   1999 counts increased to 2009 total approach volume of CDTC STEP Model

Leg	Flow (veh/hr)	Avg Delay	Avg Queue	Max Delay	Max Queue
Washington Ave EB	1454	14 sec	5	25 sec	9
Fuller Rd NB	709	10 sec	2	18 sec	3
Washington Ave WB	1642	16 sec	7	30 sec	13
Fuller Rd SB	1062	6 sec	2	11 sec	3

VMT was estimated using a quarter mile approach for each leg of the intersection. Speeds were calculated over that same distance as 15 mph under existing conditions and 29 mph with the roundabout.

The STEP model was also used to calculate seconds of delay for vehicles with the existing signalized intersection for the no-build scenario. The NYSDOT Roundabout Design Unit conducted an analysis of the proposed improvement using the RODEL Roundabout Capacity model to calculate seconds of delay for approach vehicles under the build scenario.

EMISSIONS

Δ VOC	- 24.17 kg/day
Δ NOx	- 1.94 kg/day
Δ CO	- 24.17 kg/day
Δ PM10	NA
Δ PM2.5	NA
Δ Total	- 26.11 kg/day (0.029 tpd)

METHODOLOGY/ASSUMPTIONS:

The NYSDOT software package CMAQtraq was used to estimate emissions, using the "Traffic Flow Improvements" module. Effects were calculated for 250 days/year with the following emissions factors (g/mile):

CO = Before: 18.01   After: 16.02
VOC = Before: 1.01   After: 0.71
NOx = Before: 0.95   After: 0.79

COSTS

	Project life:__NA__ yrs	Interest rate: ___7__%

	CMAQ	NON-CMAQ	TOTAL	METHODOLOGY/ASSUMPTIONS:
Capital	$2.0 M	$2.87 M	$4.87 M	
Adm/oper	$0	$0	$0	Funding will include planning, design, and construction of the intersection improvement. A cost effectiveness calculation was not provided by the project sponsor.
Total	$2.0 M	$2.87 M	$4.87 M	
Total annualized public cost:	$467,981			
Annual revenues:	None			
Net public cost:	$4.87 M			
Annual private cost	NA			
Total net cost	$4.87 M			

Category: **TRAFFIC FLOW IMPROVEMENTS**		Subcategory: **Freeway Management**
CMAQ Project ID: LA20030008		Project Year: 2003
Location: Baton Rouge, Louisiana		MPO: Capital Regional Planning Commission

Description: **ITS on Interstate 10 from Acadian St. to Highland Blvd.** - Project will continue phase II of the Baton Rouge ITS plan, and include installing freeway ITS components along I10 from Acadian St. to Highland Blvd. to assist with incident detection and response, motorist assistance, and surveillance.

TRAVEL IMPACTS

		METHODOLOGY/ASSUMPTIONS:
Δ Vehicle trips:	NA	
Δ VMT:	NA	
Δ Speed:	NA	The overall level of VMT and vehicle trips is not assumed to be affected.
Δ Delay:	NA	Emissions reductions will occur through a reduction in nonrecurring congestion.
Δ SOV	NA	
Δ CP/VP	NA	
Δ Transit	NA	
Δ Walk	NA	
Δ Bike	NA	

EMISSIONS

		METHODOLOGY/ASSUMPTIONS:
Δ VOC	- 189.601 kg/day	Emissions factors for baton rouge based on MOBILE Model; assumed running speed of 40 MPH. Emissions reductions were applied to the length of I10, as follows:
Δ NOₓ	- 488.972 kg/day	1) Freeway emissions = freeway VMT (from Tranplan model) * Emissions factor (from MOBILE in grams/mile)
Δ CO	NA	2) Freeway emissions due to nonrecurring congestion = freeway emissions * 0.049 (assumes 4.9% of freeway emissions are caused by nonrecurring congestion using data from Lindley, J. A. "Urban Freeway Congestion: Quantification of the Problem and Effectiveness of Potential Solutions." 1987.)
Δ PM₁₀	NA	
Δ PM₂.₅	NA	
Δ Total	-678.573 kg/day (0.748 tpd)	3) Emissions reduced due to program = freeway emissions due to nonrecurring congestion * effectiveness factor. Effectiveness factor assumed to be 0.90, based on effectiveness rate of 50% for Incident Detection and Response, 25% for Motorist Assistance, and 15% for Surveillance.

COSTS

				Project life: __ 10 _ yrs	Interest rate: __ 7 ___%

	CMAQ	NON-CMAQ	TOTAL	Methodology/Assumptions:
Capital	$2,712,940	$0	$2,712,940	Assumes that project has benefits 260 days per year (all weekdays). Cost-effectiveness calculation will need to take into account the life of the capital equipment.
Adm/oper	$0	$0	$0	
Total	$2,712,940	$0	$2,712,940	
Total annualized public cost:	$443,109			
Annual revenues:	None			
Net public cost:	$2,712,940			
Annual private cost	NA			
Total net cost	$2,712,940			

NOTE: Assumption that 4.9% of freeway emissions are due to nonrecurring congestion is based on old source and there may be more recent data available. Calculation seems to assume 90% reduction in emissions associated with non-recurrent congestion, and it is not clear that the effectiveness of each ITS component should be additive.

Category: **TRAFFIC FLOW IMPROVEMENTS**			Subcategory: **Freeway Management**		
CMAQ Project ID: WA20040027			Project Year: 2004		
Location: Seattle, Washington			MPO: Puget Sound Regional Council		
Description: **Duwamish ITS System** - This project will minimize the conflicts among freight movement, transit travel, commuter traffic, and ferry access, while enhancing safety and mobility for people and goods. The project will include, among other things: interconnection of traffic signals and controller equipment upgrading, installation of variable message signs and other driver information systems, implementation of traffic control strategies, and CCTV and roadway signs to monitor traffic conditions and accidents.					
TRAVEL IMPACTS					
Δ Vehicle trips:	NA	METHODOLOGY/ASSUMPTIONS: The assumptions for input into TCM Tools included the expectation that the project will improve both peak and off-peak period speeds by 10% (from 19 to 21 mph), with an average daily traffic (ADT) of 200,000 in 2010.			
Δ VMT:	NA				
Δ Speed:	+ 2 mph				
Δ Delay:	NA				
Δ SOV	NA				
Δ CP/VP	NA				
Δ Transit	NA				
Δ Walk	NA				
Δ Bike	NA				
EMISSIONS					
Δ VOC	- 76 kg/day	METHODOLOGY/ASSUMPTIONS: Emissions reductions calculated using the TCM Tools program created by Parsons Brinkerhoff and Sierra Research in 1994, which applies project data to the project year's (2004) MOBILE emissions factors and regional data to produce the emissions reductions for CO, VOCs, and NOx.			
Δ NO$_X$	- 4 kg/day				
Δ CO	- 939 kg/day				
Δ PM$_{10}$	NA				
Δ PM$_{2.5}$	NA				
Δ Total	- 80 (kg/day) (0.088 tpd)				

COSTS

				Project life:__10__ yrs	Interest rate:__7___%
	CMAQ	NON-CMAQ	TOTAL	METHODOLOGY/ASSUMPTIONS:	
Capital	$998,037	$1,001,963	$2.0 M	Total Project Cost: ~$2,000,000 (other funds in the project include other Federal and State/local funds). Cost-effectiveness was not provided by the project sponsor.	
Adm/oper	$0	$0	$0		
Total	$998,037	$1,001,963	$2.0 M		
Total annualized public cost:	$318,190			Project assumes benefits 252 days per year.	
Annual revenues:	None				
Net public cost:	$2.0 M				
Annual private cost	NA				
Total net cost	$2.0 M				

NOTE: The CMAQ project amount in the CMAQ database is $862,192 for this project. Total project costs and CMAQ funding provided by the State were used to calculate the amounts in the table.

Category: **TRAFFIC FLOW IMPROVEMENTS**	Subcategory: **Freeway Management**

CMAQ Project ID: CT20050001	Project Year: 2005
Location: Connecticut	MPO: South Central Regional COG MPO

Description: **Incident Management System on I-95** - This project will fund the construction of a 13.94 mile portion of an incident management system on I-95 from exit 56 to exit 64. The overall project will include the installation of a fiber-optic communication system, video surveillance, traffic flow monitors, and a link to the Bridgeport Operations Center. The incident management project will provide an effective means of managing traffic congestion by allowing operational problems to be identified sooner and by enabling faster dispatch of the proper response equipment and medical services to a site.

TRAVEL IMPACTS

		METHODOLOGY/ASSUMPTIONS:
△Vehicle trips:	NA	Length is 13.94 miles, 10.22 in Fairfield County and 3.72 miles in Greater Connecticut.
△VMT:	- 23,561 /day	
△Speed:	NA	The "Connecticut Freeway Management System" report documents the effects of incident management systems. Based on the report, this type of system will result in annual delay savings of 1.72 million vehicle hours (MVH) for a corridor length of 65 miles, based on a congested incident speed of 5 mph, and a free flow speed of 55 mph. The daily VMT traveled without an IM system in place is 1.72 MVH x 5 mph / 365 days = 23,561 VMT.
△Delay:	NA	
△SOV	NA	
△CP/VP	NA	
△Transit	NA	
△Walk	NA	
△Bike	NA	

EMISSIONS

		METHODOLOGY/ASSUMPTIONS:
△ VOC	- 6.11 kg/day	For Fairfield County:
△ NOx	- 3.00 kg/day	VOC reduction = (1.700 – 0.490) x 23,561 / 65 miles x 10.22 miles = 4.48 kg/day
△ CO	NA	NOx reduction = (1.988 – 1.395) x 23,561 / 65 miles x 10.22 miles = 2.20 kg/day
△ PM$_{10}$	NA	PM2.5 reduction = (0.288 – 0.287) x 23,561 / 65 miles x 10.22 = 0.004 kg/day
△ PM$_{2.5}$	- 0.004 kg/day	
△Total	- 9.11 kg/day (0.01 tpd)	For Greater Connecticut Area: VOC reduction = (1.700 – 0.490) x 23,561 / 65 miles x 3.72 miles = 1.63 kg/day NOx reduction = (1.988 – 1.395) x 23,561 / 65 miles x 3.72 miles = 0.80 kg/day

COSTS

Annualized public costs				Project life:_10__ yrs	Interest rate: ___7__%

	CMAQ	NON-CMAQ	TOTAL	METHODOLOGY/ASSUMPTIONS:
Capital	$1,279,246	$142,138	$1,421,384	
Adm/oper	$0	$0	$0	
Total	$1,279,246	$142,138	$1,421,384	
Total annualized public cost:	$218,725			
Annual revenues:	None			
Net public cost:	$1,421,384			
Annual private cost	NA			
Total net cost	$1,421,384			

NOTE: Emissions reductions reported in CMAQ database differ from estimates provided or calculated from sponsor-provided documentation (-9.1 kg/day VOC, -4.6 kg/day NOx).

Category: **TRAFFIC FLOW IMPROVEMENTS**	Subcategory: **Freeway Management**

CMAQ Project ID: Not Yet Assigned	Project Year: 2007
Location: Birmingham, AL	MPO: Birmingham RPC

Description: **Alabama Service Patrols Program** - The Alabama Service and Assistance Patrol, or "A.S.A.P" has been a program of the Alabama Department of Transportation and the Alabama State Troopers since 1997. A.S.A.P service trucks offer a variety of free services to disabled motorist to reduce response time by appropriate authorities responding to traffic related incidents and to minimize major disruption of interstate flow at an incident location. In addition, video cameras placed along interstate routes permit the State Troopers to monitor traffic flow at priority, high-traffic flow locations, which are more likely to have a traffic incident. A.S.A.P. operators patrol from 6 am to 10 pm.

TRAVEL IMPACTS

		METHODOLOGY/ASSUMPTIONS:
Δ Vehicle trips:	NA	306 total vehicles were relocated to ramps and 69 accidents were relocated from a travel lane. Data provided by Alabama DOT for 7/3/2006-6/29/2007 period.
Δ VMT:	NA	Estimated percentage of disabled vehicles which occur during peak period = 25%
Δ Speed:	NA	Estimated percentage of incidents which occur in peak period = 50%
Δ Delay: (vehicle hours/incident)	- 3,849 / incident	Total numbers of accidents (travel lane opened during project) = 111 accidents
		Traffic volume prior to project = 1400 vehicle/hour/lane
Δ SOV	NA	Average number of blocked lanes during incidents = 1.1 lanes
Δ CP/VP	NA	Average number of lanes for the InterState highway = 3 lanes
Δ Transit	NA	Incident duration prior to project = 1.10 hours
Δ Walk	NA	Incident duration after project implementation = 0.71 hours
Δ Bike	NA	Incident Delay = Traffic volume * (Average number of blocked lanes during incidents / total lanes in corridor) * Incident duration
		Change in delay = Incident delay without project – Incident delay with project (7,126 vehicle hours – 3,277 vehicle hours = 3,849 vehicle hours)

EMISSIONS

		METHODOLOGY/ASSUMPTIONS:
Δ VOC	- 31.25 kg/day	HC Idle Emissions Factor during incident 19.018 grams/hour
Δ NOx	- 11.88 kg/day	NOx Idle Emissions Factor during incident 7.230 grams/hour
Δ CO	NA	PM 2.5 Standard PM idle emissions factor 0.072 grams/hour (2005)
Δ PM$_{10}$	NA	PM 2.5 Standard NOx idle emissions factor 6.618 grams/hour (2005)
Δ PM$_{2.5}$	- 0.12 kg/day	
Δ Total	-43.1 kg/day	For each pollutant, the Change in delay * Emissions Factor / 1,000 * 111 annual incidents / 260 working days = kg of emissions reduced per day.

COSTS

Annualized public costs				Project life:__1__ yrs	Interest rate: ___7__ %

	CMAQ	NON-CMAQ	TOTAL	METHODOLOGY/ASSUMPTIONS:
Capital	$0	$0	$0	Cost information is provided for 1 year of operating funding, or 260 days per year.
Adm/oper	$240,000	$560,000	$800,000	
Total	$240,000	$560,000	$800,000	Project sponsor calculated cost effectiveness as: the Annual project cost / (Emissions reduced * 260 days)
Total annualized public cost:		$800,000		
Annual revenues:		$800,000		VOC Cost Effectiveness: $105 dollars/kg/year
Net public cost:		$800,000		NOx Cost Effectiveness: $277 dollars/kg/year
Annual private cost		NA		PM2.5 Cost Effectiveness: $27,827 dollars/kg/year
Total net cost		$800,000		

Category: **TRAFFIC FLOW IMPROVMENTS**		Subcategory: **HOV Lanes**
CMAQ Project ID: TX20020069	Project Year: 2002	
Location: Dallas, Texas	MPO: North Central Texas Council of Governments	

Description: **Dallas HOV Interchange** - This project will fund the construction of an HOV Interchange at IH635 and US75 in Dallas. The project was selected by the Regional Transportation Council (Dallas-Fort Worth MPO policy body) in 2001. The project was originally funded almost entirely through the National Highway System (NHS) program, but in 2001, CMAQ funding was added for the construction of the HOV portion of the interchange. The project was selected during a strategic assessment of regional priorities by the Regional Transportation Council. The reduction methodology is adapted from "The Texas Guide to Accepted Mobile Source Emissions Reduction Strategies" published by Texas Transportation Institute, 2003.

TRAVEL IMPACTS

		METHODOLOGY/ASSUMPTIONS:
Δ Vehicle trips:	- 2,929 /day	Assume the number of HOV users per day is 10053.72 and the average vehicle occupancy of rideshares is 2.14 persons per vehicle. Also assume the percentage of people attracted to the HOV which:
Δ VMT:	- 58,589 /day	
Δ Speed:	NA	Use transit = 0.14
Δ Delay:	NA	Use transit and previously drove alone= 0.56
Δ SOV	NA	Use ride share = 0.83
Δ CP/VP	NA	Use ride share and previously drove alone = 0.56
Δ Transit	NA	Calculate daily vehicle trip reduction = 10053.72 users * (0.14 * 0.56 + 0.83 * 0.56) * (1 – 1 / 2.14 persons/vehicle).
Δ Walk	NA	
Δ Bike	NA	Calculate VMT reduction = 2,929 trips reduced * 20 mile average auto trip length.

EMISSIONS

		METHODOLOGY/ASSUMPTIONS:
Δ VOC	- 68.78 kg/day	Emissions reductions calculated using Mobile6 running emissions factors, assuming a 43 mph running speed on freeways before the project = NOx: 1.22 and VOC 0.53 grams/mile. Speed-based running exhaust emissions factor for the HOV facility, assuming a 51 mph speed = NOx: 1.32 and VOC 0.51 grams/mile. Speed-based running exhaust emissions factor for general purpose lanes, assuming a 43 mph speed = NOx: 1.22 and VOC 0.53 grams/mile.
Δ NO$_X$	- 135.32 kg/day	
Δ CO	Kg	
Δ PM$_{10}$	NA	Calculate change in running exhaust emissions from vehicles shifting from general purpose lanes to HOV lanes. Assume Average Peak Traffic on HOV lanes after project is 783 vehicles/hour and 6 peak hours per day. The HOV length is 20.9 miles. Calculate change in running exhaust emissions from vehicles in general purpose lanes as a result of vehicles shifted away from general purpose lanes. Assume Average Peak Traffic on general purpose lane before project is 10,358 vehicles/hour and the Average Peak Traffic on general purpose lane after project is 10,797 vehicles/hour.
Δ PM$_{2.5}$	NA	
Δ Total	- 204.10 kg/day) (0.22 tpd)	
		Calculate the reduction in auto start emissions from trip reductions using an auto trip end emissions factor for NOx: 0.39 grams/mile and VOC: 1.25 grams/mile and the trip reductions. Calculate the reduction in auto running exhaust emissions from trip reductions using the emissions factor before project implementation and the vehicle miles reduced.

COSTS

Project life: 20 yrs   Interest rate: ___7___ %

	CMAQ	NON-CMAQ	TOTAL	METHODOLOGY/ASSUMPTIONS:
Capital	$17.152 M	$237,418,093	$254,570,093	
Adm/oper	$0	$0	$0	The funding details for this project are as follows:
Total	$17.152 M	$237,418,093	$254,570,093	$254,570,093 total at letting ($229,853,137 Category 2 funds (NHS account in Texas), $17,152,000 CMAQ, $4,288,000 State funds, $122,856 TxDOT Green Ribbon Funds, and $3,154,100 local). A cost effectiveness calculation was not provided by the project sponsor.
Total annualized public cost:		$28,194,000		
Annual revenues:		None		
Net public cost:		$254,570,093		
Annual private cost		NA		
Total net cost		$254,570,093		

Category: **SHARED RIDE PROGRAMS**			Subcategory: **Regional Ridesharing**		
CMAQ Project ID: MD20020010			Project Year: 2002		
Location: Maryland			MPO: No MPO Identified – State-sponsored Project		
Description: **11 County Ridesharing Program Operations** - This project will promote and encourage the establishment of carpools and vanpools in eleven Maryland Ridesharing Programs. The programs are operated by Anne Arundel County, Baltimore Metropolitan Council, Calvert Frederick, Harford, Howard, Montgomery and Prince George's counties AND Tri-County Council.					

TRAVEL IMPACTS

Δ Vehicle trips:	- 3,000 /day	METHODOLOGY/ASSUMPTIONS:
Δ VMT:	- 84,000 /day	Assumes 12,360 individual rideshare applicants in the eleven programs (based on actual data from past years).
Δ Speed:	NA	
Δ Delay:	NA	Vehicle trips reduced = 12,360 applicants * 0.24 formation rate = 3,000 vehicle trips reduced per day (assumes 24 percent of total applicants will take part in ridesharing each day).
Δ SOV	NA	
Δ CP/VP	NA	
Δ Transit	NA	
Δ Walk	NA	VMT reduced = 3,000 vehicle trips * 28 miles = 84,000 vehicle miles per day (assumes one-way trip is 14 miles).
Δ Bike	NA	

Emissions

Δ VOC	- 35.0 kg/day	Methodology/Assumptions:
Δ NO$_x$	- 110.0 kg/day	Emissions reductions were calculated by multiplying VMT reduction by per-mile emissions factors. Emissions were calculated based on 2005 stabilized running emissions factors developed for Baltimore region based on MOBILE Model (11/16/1999 run). The assumed running speed is 35 MPH.
Δ CO	NA	
Δ PM$_{10}$	NA	
Δ PM$_{2.5}$	NA	
Δ Total	- 145 kg/day (0.16 tpd)	VOC Emissions Factor: 0.4 grams/mile
		NOx Emissions Factor: 1.3 grams/mile
		Note: emissions factor assumes no cold start or hot soak emissions are affected.

COSTS

			Project life:__ 1 _ yrs	Interest rate: __ 7 __%	
	CMAQ	NON-CMAQ	TOTAL	METHODOLOGY/ASSUMPTIONS:	
Capital	$0	$0	$0	This project has been funded with CMAQ over multiple years. All calculations apply to 1 year of operating costs and emissions reductions.	
Adm/oper	$956,000	$0	$956,000		
Total	$956,000	$0	$956,000		
Total annualized public cost:	$956,000				
Annual revenues:	None				
Net public cost:	$956,000				
Annual private cost	NA				
Total net cost	$956,000				

NOTE: Methodology assumes that all new carpoolers/vanpoolers previously drove alone (this may be reasonable, but perhaps should be confirmed through surveys). Carpool/vanpool formation rate should account for the fact that all new carpools/vanpools may not operate each day (this may be imbedded in the calculation of this rate).

Category: **SHARED RIDE PROGRAMS**			Subcategory: **Regional Ridesharing**		

CMAQ Project ID: PA20050202	Project Year: 2005
Location: Oakland, Pennsylvania	MPO: Southwestern Pennsylvania Commission MPO

Description: University of Pittsburgh TDM Program - The Oakland Area TDM Program will fund the operation, marketing, and administration of a ridesharing program for employees and employers in Oakland, part of metropolitan Pittsburgh. Sponsored by the University of Pittsburgh, the program expanded existing ridesharing coordination, employer-sponsored vanpools, and carpool programs. The promotional budget will include 20% print media, 20% signage, 30% radio, and 30% promotional brochures.

TRAVEL IMPACTS

Δ Vehicle trips:	- 2,024 /day	METHODOLOGY/ASSUMPTIONS:
ΔVMT:	- 22,062 /day	Annual Program Budget = $300,000 - $50,000 overhead Average work trip length = 10.9 miles Estimate the number of vehicle trips reduced per $1 of the program = 2.04 trips day
Δ Speed:	NA	
Δ Delay:	NA	Vehicle Trip Reduction = ($250,000 * 2.04) / 252 = - 2,024 trips per day
Δ SOV	NA	Round trip VMT Reduction = -2,024 * 10.9 miles = -22,062 miles per day
Δ CP/VP	NA	
ΔTransit	NA	
Δ Walk	NA	
Δ Bike	NA	

EMISSIONS

Δ VOC	- 26.2 kg/day	METHODOLOGY/ASSUMPTIONS:
Δ NOx	- 30.9 kg/day	Total VMT was distributed into Rural and Urban locations and Freeway, Arterial, and Local facility types using data from a 1996 County Percent VMT by Facility and Area Type document. Emissions reductions calculated using Mobile 5a emissions factors for each of these locations and facility types.
Δ CO	-187.4 kg/day	
Δ PM10	NA	
Δ PM2.5	NA	
ΔTotal	-57.1 kg/day (0.063 tpd)	

COSTS

Project life:__2__ yrs   Interest rate: ___7__%

	CMAQ	NON-CMAQ	TOTAL	METHODOLOGY/ASSUMPTIONS:
Capital	$0	$0	$0	Funding is for 2 years of operating subsidy for the program
Adm/oper	$480,000	$120,000	$600,000	
Total	$480,000	$120,000	$600,000	
Total annualized public cost:	$358,670			
Annual revenues:	None			
Net public cost:	$600,000			
Annual private cost	NA			
Total net cost	$1.08 M			

NOTE: The documentation is unclear on details about the cost assumptions. The travel methodology suggests that the program budget on which the trip reduction benefits are estimated is $250,000/year, though this is supposed to be the budget without overhead; for this project. Moreover, the effectiveness calculation is based on an assumption of 2.04 vehicle trips reduced per day for each $1 of the program, but documentation was not provided; it is unclear where this metric was derived and whether it is appropriate for this program. .

Category: **SHARED RIDE PROGRAMS**	Subcategory: **Regional Ridesharing**

CMAQ Project ID: Not Yet Assigned	Project Year: 2007
Location: Birmingham, AL	MPO: Birmingham Regional Planning Commission

Description: **CommuteSmart Commuter Services Program Operations** - The project will fund the continuing operation of the CommuteSmart Commuter Services Program in Birmingham, Alabama. The program includes a ridesharing database, a vanpool program with up to 34 vans in 2007 and a carpool program.

TRAVEL IMPACTS

		METHODOLOGY/ASSUMPTIONS:
Δ Vehicle trips:	- 311.78 /day	Number of Vanpool vans = 34 vehicles
Δ VMT:	- 9,469.98 /day	Average van occupancy = 9.64 people per van
Δ Speed:		Estimated percent of vanpoolers previously took carpools = 9%
Δ Delay:		Annual Van trips = 17,380 trips/year
Δ SOV		Annual Van miles = 1,078,692 miles/year
Δ CP/VP	- 76 carpool trips/day	Annual Passenger Trips = 107,303 trips/year
		Annual Passenger Miles = 4,241,282 miles/year
		Passenger trip length per trip (one way) = 39.53 miles per trip
Δ Transit		Average auto occupancy = 1.09 people per car
Δ Walk		Number of days project affected per year = 260 days per year
Δ Bike		

Daily Vehicle Trip Reduction: (107,303 passenger trips / 1.09 average auto occupancy – 17,380 van trips) / 260 days/year = 311.78 daily vehicle trip reduction.
Of those trips, carpool trip reduction = 76 trips/day.

VMT reduction (taking into account the van miles): 4,241,282 passenger miles / 1.09 auto occupancy x ((1 – 9% percent of vanpoolers previously took carpools) - 1,078,692 van miles) / 260 days/year = 9,469.98 daily VMT.
Of that VMT reduction, carpool VMT reduction = 188,929 miles/year / 260 days/year = 726.65 daily carpool VMT reduction.

EMISSIONS

Δ VOC	- 10.21 kg/day	METHODOLOGY/ASSUMPTIONS:
Δ NOx	- 11.96 kg/day	Emissions reductions calculated using Mobile6 emissions factors for 2005 at a 35 mph average operating speed.
Δ CO	NA	Auto HC emissions factor 1.1640 grams/mile
Δ PM10	NA	Auto NOx emissions factor 1.2720 grams/mile
Δ PM2.5	- 0.133 kg/day	Van HC emissions factor 1.5630 grams/mile
Δ Total	-22.17 kg/day (0.024 tpd)	Van NOx emissions factor 1.5160 grams/mile
		Auto PM2.5 emissions factor, 0.0133 grams/mile
		Van PM2.5 emissions factor, 0.0140 grams/mile

COSTS

Project life:__1__ yrs   Interest rate: ___7__ %

	CMAQ	NON-CMAQ	TOTAL	METHODOLOGY/ASSUMPTIONS:
Capital	$0	$0	$0	Total project cost = 1.07 Capital recovery factor * $700,000 = $749,000 / year
Adm/oper	$700,000	$0	$700,000	
Total	$700,000	$0	$700,000	Cost Effectiveness Calculation: $749,000 project annual cost / (Emissions reduced (kg/day) * 260 days of effect)
Total annualized public cost:	$700,000			
Annual revenues:	None			HC Cost Effectiveness = $282 per kg/year
Net public cost:	$700,000			NOx Cost Effectiveness = $241 per kg/year
Annual private cost	NA			PM2.5 Cost Effectiveness = $21,707 per kg/year
Total net cost	$700,000			

Category: **SHARED RIDE PROGRAMS**				Subcategory: **Vanpool Programs**

CMAQ Project ID: UT20020006	Project Year: 2002
Location: Salt Lake and Ogden, Utah	MPO: Wasatch Front Regional Council

Description: **15 New Vans for Vanpool Leasing Program** - This project is the purchase of 15 8-passenger vans for Salt Lake City and Ogden areas.

TRAVEL IMPACTS

		METHODOLOGY/ASSUMPTIONS:
△ Vehicle trips:	Na	Assumes 8 passengers per van and 46 miles average daily round trip.
△ VMT:	- 5,520 /day	
△ Speed:	NA	Daily VMT reduction = 8 passengers/van x 15 vans x 23 miles one-way trip x 2 trip lengths reduce reduced/day = 5,520 VMT reduced.
△ Delay:	NA	
△ SOV	NA	
△ CP/VP	NA	
△ Transit	+ 15 vans	
△ Walk	NA	
△ Bike	NA	

EMISSIONS

		METHODOLOGY/ASSUMPTIONS:
△ VOC	- 12.2 kg/day	Assumes that project has benefits 250 days/year.
△ NO_x	- 14.9 kg/day	Emissions reductions calculated by applying passenger car CO, NOx, and VOC g/mile rates for freeways and arterials.
△ CO	- 136.9 kg/day	
△ PM_{10}	NA	
△ $PM_{2.5}$	NA	
△ Total	- 27.1 kg/day (0.03 tpd)	

COSTS

				Project life:_5___ yrs	Interest rate: ___7__%

	CMAQ	NON-CMAQ	TOTAL	METHODOLOGY/ASSUMPTIONS:
Capital	$448,000	$0	$448,000	Cost benefit was calculated using a weighted ranking system that considered the following:
Adm/oper	$0	$0	$0	1) (10%) Project in a congested corridor
Total	$448,000	$0	$448,000	2) (15%) Length (years) of project effectiveness
Total annualized public cost:		$128,200		3) (25%) Emissions reduction
				4) (25%) Congestion reduction (VHT)
Annual revenues:		None		5) (25%) Cost
Net public cost:		$448,000		This objective ranking was then combined with subjective rankings by staff and 3 different committees consisting of city planners and elected officials.
Annual private cost		NA		
Total net cost		$448,000		

NOTE: The project description provided by the sponsor lists a purchase of 15 vehicles, while the project description in the CMAQ database describes this project as the purchase of 70 vans. The project cost provided by the sponsor lists the project cost as $377,582, while the database lists the CMAQ-allotted funds as $448,000. The discrepancy might stem from the difference in the amount of vehicles purchased. The calculation does not explicitly account for any increase in emissions from the vans operating.

Category: **SHARED RIDE PROGRAMS**		Subcategory: **Vanpool Program**
CMAQ Project ID: UT20050005		Project Year: 2005
Location: Ogden and Layton, Utah		MPO: Wasatch Front Regional Council MPO
Description: **5 New Vans for Vanpool Leasing Program** - This project is the expansion of the UTA Vanpool Leasing Program through the purchase of 5 vans for the Ogden and Layton area.		

TRAVEL IMPACTS		
Δ Vehicle trips:	NA	**METHODOLOGY/ASSUMPTIONS:**
Δ VMT:	- 3,000 /day	Daily VMT reduction = 5 vans x 8 passengers per van x 45 mile one-way trip x 2
Δ Speed:	NA	trip lengths reduced/day / 1.2 personal auto occupancy rate = 3,000 VMT reduced.
Δ Delay:	NA	
Δ SOV	NA	
Δ CP/VP	+ 5 vans	
Δ Transit	NA	
Δ Walk	NA	
Δ Bike	NA	

EMISSIONS		
Δ VOC	- 3.2 kg/day	**METHODOLOGY/ASSUMPTIONS:**
Δ NO$_x$	- 4.0 kg/day	Passenger car CO, NOx, VOC g/mile rates for freeways and arterials applied.
Δ CO	- 37.2 kg/day	*Freeway:*
Δ PM$_{10}$	NA	VOC emissions factor = 1.38 g/mile
Δ PM$_{2.5}$	NA	NOx emissions factor = 1.98 g/mile
Δ Total	- 7.2 kg/day (0.008 tpd)	CO emissions factor = 14.32 g/mile *Arterial:* VOC emissions factor = 2.32 g/mile NOx emissions factor = 2.03 g/mile CO emissions factor = 29.61 g/mile

COSTS				
			Project life:____ yrs	Interest rate: ___7__%
	CMAQ	NON-CMAQ	TOTAL	**METHODOLOGY/ASSUMPTIONS:**
Capital	$148,000	$32,866	$180,866	Cost/benefit calculation – Projects were ranked from first to last for cost/congestion benefit (VHT reduced) and cost/air quality benefit (tons emissions). The benefits were multiplied by the project life – the number of years the benefits would be returned. The two rank scores were then added together and all projects were ranked again based on this composite score. The objective ranking was then combined with subjective rankings by staff and 3 different committees consisting of city planners and elected officials.
Adm/oper	$0	$0	$0	
Total	$148,000	$32,866	$180,866	
Total annualized public cost:	$47,676			
Annual revenues:	None			
Net public cost:	$180,866			
Annual private cost	NA			
Total net cost	$180,866			

NOTE: Emissions reductions provided by the State sponsor do not match those reported in the CMAQ database (-5 VOC, -54 CO, and -6 NOx). The calculation does not explicitly account for any increase in emissions from the vans operating..

Category: **SHARED RIDE PROGRAMS**	Subcategory: **Vanpool Program**

CMAQ Project ID: KY20060004	Project Year: 2006
Location: Lexington, Kentucky	MPO: Lexington Area MPO

Description: **6 New Vans for LexTran Vanpool Service** - This project is the purchase of six new 12-passenger vans for LexVan, a commuter vanpool program managed by the Lexington Bluegrass Mobility Office. LexVan leases these passenger vans to groups of people who vanpool to work. The passengers are matched to a vanpool group using ridesharing computer software and each passenger pays a monthly fare which covers all operating costs, fuel, and insurance. This program has a direct effect in reducing the number of single occupant vehicles (SOVs) during peak hours.

TRAVEL IMPACTS

		METHODOLOGY/ASSUMPTIONS:
Δ Vehicle trips:	- 132 /day	Each new van removes a maximum of 11 SOVs from the road system and has a 50 mile average LexVan round trip.
Δ VMT:	- 3,300 /day	
Δ Speed:	NA	
Δ Delay:	NA	Vehicle trip reduction = 6 vans x 11 SOV removal x 2 trips = 132 vehicle trips removed per day.
Δ SOV	NA	
Δ CP/VP	+ 6 vans/day	
Δ Transit	NA	VMT reduction = 132 vehicle trips removed x 25 miles per trip (based on 50 mile round trip) = 3,300 VMT reduction/day.
Δ Walk	NA	
Δ Bike	NA	

EMISSIONS

		METHODOLOGY/ASSUMPTIONS:
Δ VOC	- 10.40 kg/day	The emissions rates are for hydrocarbons (HC), carbon monoxide (CO), and oxides of nitrogen (NOx) and are from U.S. Environmental Protection Agency (EPA) highway vehicle emissions factor models. They assume an average properly maintained vehicle on the road in July, operating on typical gasoline on a warm summer day (72-96 degrees F).
Δ NO$_x$	- 5.28 kg/day	
Δ CO	- 80.19 kg/day	
Δ PM$_{10}$	NA	
Δ PM$_{2.5}$	NA	
Δ Total	- 15.68 kg/day (0.017 tpd)	VOC: 3.15 g/mile x 50 mile vanpool round trip x 66 SOV reduction / 1000 = 10.40 kg/day
		NOx: 1.60 g/mile x 50 mile vanpool round trip x 66 SOV reduction / 1000 = 5.28 kg/day
		CO: 24.30 g/mile x 50 mile vanpool round trip x 66 SOV reduction / 1000 = 80 .19 kg/day

COSTS

Project life:__ 5 _ yrs	Interest rate: __ 7 __%

	CMAQ	NON-CMAQ	TOTAL	METHODOLOGY/ASSUMPTIONS:
Capital	$96,000	$24,000	$120,000	The per day reductions based on an average of 21 work/school days in a given month or 252 days in the 12 months per year period used for the emissions analysis.
Adm/oper	$0	$0	$0	
Total	$96,000	$24,000	$120,000	
Total annualized public cost:	$30,643			The cost for the purchase of 6 new 12-passenger vans is estimated at $20,000 each. This includes the installation of extra equipment. Such as side steps, striping, grab handles, etc. The entire local match is paid from the LexVan (vanpool) program fares.
Annual revenues:	None			
Net public cost:	$120,000			
Annual private cost:	NA			
Total net cost:	$120,000			

NOTE: The calculation does not explicitly account for any increase in emissions from the vans operating.

Category: **SHARED RIDE PROGRAMS**				Subcategory: **Park and Ride Lots**
CMAQ Project ID: MD20000017			Project Year: 2000	
Location: Maryland			MPO: Baltimore Metropolitan Council	
Description: **Two New 25-Space Lots** - Construction of two new park and ride facilities at I-95 interchanges at MD 272 and MD 279. Each park and ride lot will contain 25 parking spaces.				

TRAVEL IMPACTS

△ Vehicle trips:	0	METHODOLOGY/ASSUMPTIONS:
△ VMT:	- 23/day	
△ Speed:	NA	Vehicle trip reduction = 50 parking spaces * 15% utilization rate * 15% new riders = 1.15 vehicle trips reduced per day (zero change in trip starts).
△ Delay:	NA	
△ SOV	NA	
△ CP/VP	NA	VMT reduction = 1.15 vehicle trips reduced * 20 mile round trip = 23 vehicle miles reduced per day.
△ Transit	NA	
△ Walk	NA	
△ Bike	NA	Lot utilization rates and the percentage of new riders were determined from surveys at existing park and ride lots.

EMISSIONS

△ VOC	- 0.012 kg/day	METHODOLOGY/ASSUMPTIONS:
△ NO$_x$	- 0.058 kg/day	Emissions reductions were calculated by multiplying VMT reduction by per-mile emissions factors. Emissions were calculated based on 1999 emissions factors developed for the Baltimore region based on the MOBILE model. Assumed running speed is 60 mph.
△ CO	NA	
△ PM$_{10}$	NA	
△ PM$_{2.5}$	NA	
△ Total	-0.070 kg/day (0.000077 tpd)	VOC Emissions Factor: 0.552 g/ml NOx Emissions Factor: 2.559 g/mi

COSTS

Project life:__ 12 _ yrs   Interest rate: __ 7 __%

	CMAQ	NON-CMAQ	TOTAL	METHODOLOGY/ASSUMPTIONS:
Capital	$132,817	$0	$132,817	Cost-effectiveness was not provided by sponsor. In order to calculate cost-effectiveness, assume the project has benefits 250 days per year for 12 years.
Adm/oper	$0	$0	$0	
Total	$132,817	$0	$132,817	
Total annualized public cost:	$12,537			
Annual revenues:	None			
Net public cost:	$132,817			
Annual private cost	NA			
Total net cost	$132,817			

NOTES: Assumptions regarding travel impacts seem very low. In particular, a 15% utilization rate means that on average only 7.5 of the 50 new spaces are utilized, and only about one person per day is assumed to reducing a vehicle trip. Also, the assumption of a 20 mile round trip (10 miles each way) for a park and ride trip sounds low, particularly given that the lots are located in Cecil County, about 20 miles from Wilmington, DE and from Aberdeen, MD, and 40 miles from close-in Baltimore suburbs. In the CMAQ database, this project appears to have been improperly listed as in the Metropolitan Washington Council of Governments' MPO. Current Maryland State Highway Administration web site shows 25 spaces at I-95 @ MD 279 (Elkton) lot and 17 spaces at I-95 @ MD 272 (Elkton) lot.

Category: **SHARED RIDE PROGRAMS**				Subcategory: **Park and Ride Lots**	
CMAQ Project ID: WI20000034, WI20000035			Project Year: 2000		
Location: Southeastern Wisconsin			MPO: Southeastern Wisconsin RPC		
Description: **Lake Geneva and Root Creek Lot** - These are two out of a group of three park and ride lots being implemented by the WisDOT District 2 office out of a group of four candidate sites recommended in the Regional Transportation Plan for Southeastern Wisconsin. The lots are designed to encourage carpooling and use of existing public transportation.					

TRAVEL IMPACTS

Δ Vehicle trips:	0	METHODOLOGY/ASSUMPTIONS:
Δ VMT:	- 3,600/day	Assumes 450 spaces constructed for all 3 lots, average occupancy of 40%, and an average one-way commute of 10 miles. Lot-specific calculations were not conducted.
Δ Speed:	NA	
Δ Delay:	NA	
Δ SOV	NA	
Δ CP/VP	NA	Vehicle trip reduction = 450 spaces * 40% utilization rate = 180 vehicle trips per day (zero change in trip starts).
Δ Transit	NA	
Δ Walk	NA	
Δ Bike	NA	VMT reduction = 180 vehicle trips reduced * 20 mile roundtrip length = 3,600 daily VMT reduced.

EMISSIONS

Δ VOC	- 1.52 kg/day	METHODOLOGY/ASSUMPTIONS:
Δ NOx	- 3.81 kg/day	
Δ CO	NA	Emissions reductions were calculated by multiplying VMT reduction by per-mile emissions factors for a typical summer day, based on MOBILE.
Δ PM$_{10}$	NA	
Δ PM$_{2.5}$	NA	
Δ Total	- 5.33 kg/day (0.0059 tpd)	Assumes speed = 35 mph.

COSTS

Project life: __ 12 _ yrs   Interest rate: __ 7 __%

	CMAQ	NON-CMAQ	TOTAL	METHODOLOGY/ASSUMPTIONS:
Capital	$48,000	$0	$48,000	CMAQ cost of $24,000 per lot was listed in documentation provided by State and FHWA CMAQ database (cost noted here is for two lots listed in CMAQ database for FY 2000). Cost-effectiveness was not provided by sponsor. In order to calculate cost-effectiveness, assume the project has benefits 250 days per year for 20 years. Cost should be scaled up to reflect full cost of all three park and ride facilities – assume $72,000 ($24,000 x 3).
Adm/oper	$0	$0	$0	
Total	$48,000	NA	$48,000	
Total annualized public cost:	NA			
Annual revenues:	None			
Net public cost:	$48,000			
Annual private cost	NA			
Total net cost	$48,000			

NOTE: Calculated emissions reductions are based on all three park and ride lots, although separate CMAQ projects were listed only for these two park and ride lots in FY 2000. Travel calculation assumes that all users of the park and ride facility previously were driving alone (no adjustment to account for share of users who previously carpooled, unless that is somehow incorporated into the utilization factor).

Category: **SHARED RIDE PROGRAMS**			Subcategory: **Park and Ride Lots**	
CMAQ Project ID: MD20020001			Project Year: 2002	
Location: Maryland			MPO: Washington Metropolitan Council of Governments	
Description: **MD210 and MD 373 500-Space Lot** - Replace/expand existing park and ride facility at MD 210 / MD 373 by adding 500 spaces.				

TRAVEL IMPACTS

		METHODOLOGY/ASSUMPTIONS:
Δ Vehicle trips:	0	
Δ VMT:	- 5,393/day	Vehicle trip reduction = 500 parking spaces * 56% utilization rate * 45% new riders = 126 vehicle trips reduced per day (zero change in trip starts).
Δ Speed:	NA	
Δ Delay:	NA	
Δ SOV	NA	
Δ CP/VP	NA	VMT reduction = 126 vehicle trips reduced * 42.8 mile round trip = 5,393 vehicle miles reduced per day.
Δ Transit	NA	
Δ Walk	NA	
Δ Bike	NA	Lot utilization rates and the percentage of new riders were determined from surveys at existing park and ride lots.

EMISSIONS

		METHODOLOGY/ASSUMPTIONS:
Δ VOC	- 1.375 kg/day	Emissions reductions were calculated by multiplying VMT reduction by per-mile emissions factors. Emissions were calculated based on 2005 (year of service opening) emissions factors developed for Baltimore region based on Mobile model. Assumed running speed is 50 mph, based on posted speed limit.
Δ NOx	- 5.889 kg/day	
Δ CO	NA	
Δ PM10	NA	
Δ PM2.5	NA	VOC Emissions Factor: 0.255 grams/mile
Δ Total	- 7.264 kg/day (0.00801 tpd)	NOx Emissions Factor: 1.092 grams/mile

COSTS

				Project life: __ 12 __ yrs	Interest rate: __ 7 __%
	CMAQ	NON-CMAQ	TOTAL	METHODOLOGY/ASSUMPTIONS:	
Capital	$1,218,831	$0	$1,218,831	Cost-effectiveness was not provided by sponsor. In order to calculate cost-effectiveness, assume the project has benefits 250 days per year for 12 years.	
Adm/oper	$0	$0	$0		
Total	$1,218,831	$0	$1,218,831		
Total annualized public cost:	$180,050				
Annual revenues:	None				
Net public cost:	NA				
Annual private cost:	NA				
Total net cost	NA				

NOTE: In the CMAQ database, this project appears to have been improperly listed as in the Pedestrian/Bicycle category. Current Maryland State Highway Administration web site shows MD 210 @ MD 373 (Aceokeek) park and ride lot contains 489 spaces.

Category: **SHARED RIDE PROGRAMS**			Subcategory: **Park-and-Ride Lots**		
CMAQ Project ID: KY20050012			Project Year: 2005		
Location: Union/Walton, Kentucky			MPO: Ohio-Kentucky-Indiana Regional COG		
Description: **Walton/Union Lot with 200 Spaces** - This project is the continued expansion and development of park-and-ride facilities in Northern Kentucky, along fixed transit routes in Boone, Kenton, and Campbell Counties. Improvements to existing lots include improving signage, adding bike parking racks, providing information kiosks, and updating Park & Ride brochures. Acquisitions of a new site for a new lot will provide approximately 200 new parking spaces for area commuters. These improvements and expansions will attract more riders to the system; thereby reducing single-occupancy automobile trips, reducing emissions, improving air quality, and reducing congestion.					

TRAVEL IMPACTS

		METHODOLOGY/ASSUMPTIONS:
Δ Vehicle trips:	0	# of parking spaces at 1 lot = 200
Δ VMT:	- 3,840 /day	Utilization rate = 60%
Δ Speed:	NA	Average round-trip distance = 32 miles
Δ Delay:	NA	
Δ SOV	NA	VMT reduction = 200 parking spaces x 60% utilization rate x 2 trip lengths reduced/day x 16 mile round trip = 3,840 vehicle miles reduced per day.
Δ CP/VP	NA	
Δ Transit	NA	
Δ Walk	NA	
Δ Bike	NA	

EMISSIONS

		METHODOLOGY/ASSUMPTIONS:
Δ VOC	- 0.88 kg/day	Emissions reductions were calculated by multiplying VMT reduction by per-mile emissions factors.
Δ NO$_x$	- 3.19 kg/day	
Δ CO	- 33.83 kg/day	Emissions factors based on local 2003 parameters, using MOBILE6 model. Assumes running speed is 41 mph (Weighted average running speed = 41 mph using OKI Travel Forecasting Model).
Δ PM$_{10}$	NA	
Δ PM$_{2.5}$	- 0.12 kg/day	
Δ Total	- 4.07 (0.0044 tpd)	VOC Emissions Factor: 0.23 grams/mile
		NOx Emissions Factor: 0.83 grams/mile
		CO Emissions Factor: 8.81 grams/mile
		PM2.5 Emissions Factor: 0.03 grams/mile

COSTS

				Project life: __ 12 _ yrs	Interest rate: __ 7 __%

	CMAQ	NON-CMAQ	TOTAL	METHODOLOGY/ASSUMPTIONS:
Capital	$844,800	$211,200	$1,056,000	Land Acquisition: $711,000 Construction: $325,000 Marketing and Outreach: $20,000
Adm/oper	$0	$0	$0	
Total	$844,800	$211,200	$1,056,000	
Total annualized public cost:	$143,695			Future operating expenses will be funded through local revenue source and fare revenue. Cost-effectiveness was not provided by the project sponsor.
Annual revenues:	None			
Net public cost:	$1,056,000			
Annual private cost	NA			
Total net cost	$1,056,000			

NOTE: The emissions reductions reported for VOC and PM2.5 in the documentation are slightly different from the amounts calculated using the methodology reported (-1 VOC, No Data PM$_{2.5}$).

Category: **SHARED RIDE PROGRAMS**	Subcategory: **Park and Ride Lots**

CMAQ Project ID: WA20010004, WA20050035	Project Year: 2005
Location: Seattle, Washington	MPO: Puget Sound Regional Council

Description: **Expansion of Terrace Station Transfer Lot to 880 Spaces** - This project will fund the construction of a multi-level parking structure on the lower-level of an existing park-and-ride lot located at I-5 and 236th Street SW.. The new garage will increase parking capacity from 388 to 880 spaces. Improvements will also be made to the elevators, pedestrian walkways, landscaping, lighting, bicycle racks, and security features. The Terrace Station park and ride primarily serves downtown Seattle and the University of Washington, for an average one-way distance of 12 miles.

TRAVEL IMPACTS

		METHODOLOGY/ASSUMPTIONS:
Δ Vehicle trips:	0	VMT reductions were calculated using a 12 mile average one-way trip distance from the lot to a final destination and the number of additional parking stalls added (492).
Δ VMT:	- 8,856 /day	
Δ Speed:	NA	Daily VMT reduction = 492 spaces x 75% utilization x 12 miles one-way trip length x 2 trips = 8,856 VMT reduction).
Δ Delay:	NA	
Δ SOV	NA	
Δ CP/VP	NA	
Δ Transit	NA	
Δ Walk	NA	
Δ Bike	NA	

EMISSIONS

		METHODOLOGY/ASSUMPTIONS:
Δ VOC	- 18.0 kg/day	Emissions reductions calculated using the TCM Tools program created by Parsons Brinkerhoff and Sierra Research in 1994, which applies project data to MOBILE emissions factors and regional data to produce the emissions reductions for CO, VOCs, and NOx.
Δ NO$_X$	- 9.0 kg/day	
Δ CO	- 145.0 kg/day	
Δ PM$_{10}$	NA	
Δ PM$_{2.5}$	NA	
Δ Total	- 27.0 kg/day (0.030 tpd)	

COSTS

				Project life:__30__ yrs	Interest rate: __7___%

	CMAQ	NON-CMAQ	TOTAL	METHODOLOGY/ASSUMPTIONS:
Capital	$4.15 M	$15.85 M	$20 M	CMAQ funding for this project was $865,000 for WA20010004 and $3,285,000 for WA20050035 in the CMAQ database. (For a total of $4.15M.) The Total Project Cost is estimated as $20,000,000. Other funds in the project include other Federal and State/local funds besides CMAQ. Cost-effectiveness was not provided by the project sponsor.
Adm/oper	$0	$0	$0	
Total	$4.15 M	$15.85 M	$20 M	
Total annualized public cost:	$1,742,000			
Annual revenues:	None			
Net public cost:	$20 M			
Annual private cost	NA			
Total net cost	$20 M			

NOTE: It is unclear how old the emissions factors used for this project are.

Category: **TRAVEL DEMAND MANAGEMENT**	Subcategory: **TDM**

CMAQ Project ID: CO20010042	Project Year: 2001
Location: Denver, Colorado	MPO: Denver Regional COG

Description: **Coordinate Telework Program** - This project funds a free telework consulting service for employers in the Denver metro area. The DRCOG's RideArrangers program provides consultations, design, implementation, evaluation, and training session assistance for interested employers.

TRAVEL IMPACTS

		METHODOLOGY/ASSUMPTIONS:
Δ Vehicle trips:	- 16,031 /week	Vehicle trip reduction = 87,127 employees at companies with a telework program x 0.05 percentage of employees that telework x 1.84 average days per week that employees telework instead of commute x 2 = 16,031 vehicle trips reduced weekly.
Δ VMT:	- 223,413 /week	
Δ Speed:	NA	
Δ Delay:	NA	
Δ SOV	NA	VMT reduction = 87,127 employees at companies with a telework program * 0.05 percentage of employees that telework * 26 mile average trip distance * 1.84 average days per week that employees telework instead of commuting = 223,413 weekly VMT reduction.
Δ CP/VP	NA	
Δ Transit	NA	
Δ Walk	NA	
Δ Bike	NA	

EMISSIONS

		METHODOLOGY/ASSUMPTIONS:
Δ VOC	- 2.0 kg/day	Emissions reductions calculated using 2006 MOBILE6 factors.
Δ NO$_X$	- 2.0 kg/day	
Δ CO	- 14.0 kg/day	
Δ PM$_{10}$	NA	
Δ PM$_{2.5}$	NA	
Δ Total	- 4 kg/day (0.0044 tpd)	

COSTS

				Project life:__ NA__ yrs Interest rate: __ 7 __%
	CMAQ	NON-CMAQ	TOTAL	METHODOLOGY/ASSUMPTIONS: Assumes 240 work days per year.
Capital	$0	$0	$0	
Adm/oper	$73,000	$18,250	$91,250	
Total	$73,000	$18,250	$91,250	
Total annualized public cost:	$91,250			
Annual revenues:	None			
Net public cost:	$91,250			
Annual private cost	NA			
Total net cost	$91,250			

Category: **TRAVEL DEMAND MANAGEMENT**	Subcategory: **TDM**

CMAQ Project ID: DC20020006, VA 20020072	Project Year: 2002
Location: District of Columbia	MPO: Metropolitan Washington COG

Description: **Employer Outreach, Bicycles** - This project provides information to businesses to encourage their employees to bike to work. Information provided to the employer would include: a list of maps and other resources; bike-on-transit and bike-to-transit information; descriptions of bicycle parking types and rack vendors; information on installing showers and lockers for employees; the name of a person or organization that would teach classes on bicycle commuting; and the names of contact people for questions on a range of subjects. The overall project and information provided will be integrated into ongoing Commuter Connection activities.

TRAVEL IMPACTS

		METHODOLOGY/ASSUMPTIONS:
Δ Vehicle trips:	- 125 /day	Assumes 7% of all employers contacted will participate in the program, with 3.5% promoting biking as a part of their voluntary program. Assumptions based on M-47c analysis assumptions.
Δ VMT:	- 500 /day	
Δ Speed:	NA	
Δ Delay:	NA	
Δ SOV	NA	3580 employers * 7% participation = 251 new employer participants.
Δ CP/VP	NA	
Δ Transit	NA	Assumes 2% of the employees at those firms will participate; 31 employees will participate. Assumes a 4 mile average trip length and 2 trips per day.
Δ Walk	NA	
Δ Bike	NA	

EMISSIONS

		METHODOLOGY/ASSUMPTIONS:
Δ VOC	- 1.0 kg/day	Used MOBILE6 emissions factors.
Δ NO$_x$	- 1.0 kg/day	Measurement of air quality impacts used modeling assumptions of measure M-47c "Employer Outreach."
Δ CO	Kg	
Δ PM$_{10}$	NA	
Δ PM$_{2.5}$	NA	
Δ Total	-2.0 kg/day (0.0022 tpd)	

COSTS

| | | | | Project life:__1__ yrs | Interest rate: ___7__% |

	CMAQ	NON-CMAQ	TOTAL	METHODOLOGY/ASSUMPTIONS: Assumes benefits 250 days/year.
Capital	$0	$0	$0	
Adm/oper	$9,000	$6,000	$15,000	
Total	$9,000	$6,000	$15,000	
Total annualized public cost:	$15,000			
Annual revenues:	None			
Net public cost:	$15,000			
Annual private cost	NA			
Total net cost	$15,000			

NOTE: Project costs in CMAQ database do not match with total costs due to how project costs are categorized by DC, Maryland, and Virginia. According to the Commuter Connections annual work program for 2002, the total cost is $15,000 ($5,000 from each jurisdiction). DC uses 100% CMAQ funds; VA uses 80% CMAQ, 20% other funds; and MD uses100% other funds.

Category: **TRAVEL DEMAND MANAGEMENT**				Subcategory: **TDM**
CMAQ Project ID: DC20050008			Project Year: 2005	
Location: District of Columbia			MPO: Metropolitan Washington COG MPO	

Description: **Guaranteed Ride Home (GRH)** - This program is an added incentive to employers and employees participating in the Commuter Connections program. It provides the security of a ride home in the event of an emergency, unscheduled overtime, or early leave departure. The program provides up to four free rides home per year in a taxi or rental car for commuters that use alternative modes of transportation at least two days per week. Since a sizeable portion of GFH applicants are already ridesharing before they apply for GFH benefits, the most common benefit of GRH may be the continuation and extension of existing ridesharing arrangement. The transportation and emissions impacts of the GRH program were measured through data from a survey conducted in the spring of 2004, which polled 1,000 commuters who had registered for GRH at some point between 2001 and 2004. The survey asked detailed questions regarding commute patterns, the permanence of mode changes, and the overall importance of the program to commuters' decisions to start/continue use of alternative modes.

TRAVEL IMPACTS

Δ Vehicle trips:	- 12,350 /day	METHODOLOGY/ASSUMPTIONS:
Δ VMT:	- 348,283/day	Based on surveys, new participants were grouped into those who work and live within the DC Metropolitan Statistical area (11,574) and those who work within the MSA but live outside (2,245). For those living within the MSA, assume 0.91 vehicle trips reduced per new participant and a 28.2 mile one-way trip length. For participants living outside the MSA, assume a 0.81 vehicle trip reduction per new participant and a 28.2 mile one-way trip length within the MSA.
Δ Speed:	NA	
Δ Delay:	NA	
Δ SOV	NA	
Δ CP/VP	NA	
Δ Transit	NA	
Δ Walk	NA	Vehicle trips reduction = (11,574 participants * 0.91 VTR per new participant) + (2,245 participants * 0.81 VTR per new participant) = 12,350 trips reduced/day.
Δ Bike	NA	
		VMT reduction = 12,350 VTR * 28.2 miles one-way trip length = 348,283 miles reduced per day.

EMISSIONS

Δ VOC	- 95.25 kg/day	METHODOLOGY/ASSUMPTIONS:
Δ NOx	- 216.82kg/day	Emissions reductions calculated using Mobile6.
Δ CO	NA	
Δ PM10	NA	
Δ PM2.5	NA	
Δ Total	- 312 kg/day	

COSTS

Project life:_____ yrs   Interest rate: ___7__%

	CMAQ	NON-CMAQ	TOTAL	METHODOLOGY/ASSUMPTIONS:
Capital	$0	$0	$0	Cost-effectiveness was not provided by the project sponsor.
Adm/oper	$772,110	$906,390	$1,678,500	
Total	$772,110	$906,390	$1,678,500	
Total annualized public cost:		$1,678,500		
Annual revenues:		None		
Net public cost:		$1,678,500		
Annual private cost		NA		
Total net cost		$1,678,500		

NOTE: Project costs in CMAQ database do not match with total costs due to how project costs are categorized by DC, Maryland, and Virginia. According to the Commuter Connections annual work program for 2005, the total cost is $1,678,500 for ($167,850 from DC; $755,325 from Maryland; $755,325 from Virginia). DC uses 100% CMAQ funds; VA uses 80% CMAQ, 20% other funds; and MD uses100% other funds. Results are based on survey data but appear to be quite large for a GRH program when considered independently from other regional TDM outreach elements that are quantified separately.

Category: **TRAVEL DEMAND MANAGEMENT**			Subcategory: **TDM**
CMAQ Project ID: RI20050010		Project Year: 2005	
Location: Rhode Island		MPO: Rhode Island State Planning Council	
Description: **Ozone Alert Days** - This program informs the public when ground level ozone will reach unhealthy levels and provides free transit service as an alternative to driving on those days. It is an effort by the State to develop public information explaining the relationship between transportation and air quality. The free transit program is only implemented on days when ground level ozone will reach unhealthy levels; in FY 2005, the program was implemented on 4 days.			

TRAVEL IMPACTS

Δ Vehicle trips:	- 2,509/day	METHODOLOGY/ASSUMPTIONS:
Δ VMT:	- 21,875/day	Utilization rates determined from existing ridership. (60,207 persons/day)
Δ Speed:	NA	
Δ Delay:	NA	Assumes 5% increase in ridership on ozone alert days. (This increase is based on a FY2004 statistically valid review). There were 4 ozone alert days in FY 2005 (60,207 persons/day * 5% = 3,010 persons/day increase for 4 days).
Δ SOV	NA	
Δ CP/VP	NA	
Δ Transit	+ 3,010 persons/day	Daily vehicle trip reduction = 3,010 persons/day ÷ 1.2 persons/vehicle = 2,509 vehicles/day reduced.
Δ Walk	NA	
Δ Bike	NA	Assumes average round trip distance = 8.72 miles
		Daily VMT reduction = 2,509 daily vehicle reduction * 8.72 miles = 21,875 daily VMT removed.

EMISSIONS

Δ VOC	- 23.0 kg/day	METHODOLOGY/ASSUMPTIONS:
Δ NOx	- 26.5 kg/day	Emissions reductions were calculated by multiplying VMT reduction by per-mile emissions factors.
Δ CO	- 251.3 kg/day	Assumes speed = 35 mph.
Δ PM10	NA	
Δ PM2.5	NA	
Δ Total	- 49.5 kg/day (0.05 tpd)	NOTE: Emissions reduction only is for 4 days.

COSTS

		Project Life: __1 year____		Interest rate: __ 7 __%
	CMAQ	Non-CMAQ	Total	METHODOLOGY/ASSUMPTIONS:
Capital	$0	$0	$0	
Adm/oper	$168,000	$0	$168,000	Calculations apply to 1 year of operating costs and emissions reductions. Note that to calculate annual emissions reductions, the daily total should be multiplied by only 4 (since effects are estimated only for episode days).
Total	$168,000	$0	$168,000	
Total annualized public cost:		$168,000		
Annual revenues:		None		
Net public cost:		$168,000		
Annual private cost		NA		
Total net cost		$168,000		

NOTE: This project is listed in CMAQ database under the "Transit" category since it largely involves transit fares.

Category: **BICYCLE/PEDESTRIAN**	Subcategory: **Bicycle Pedestrian**
CMAQ Project ID: MA20020040	Project Year: 2002
Location: Swansea, Massachusetts	MPO: Southeastern Regional Planning and Economic Development District

Description: **8.3 mile Swansea Bikeway Facility** - The Swansea bike path project forms an essential part of the future link between the Taunton River Trail and the East Bay Trail in Rhode Island. The proposed route along Old Warren Rd. is primarily a bike facility located on streets, with a few bicycle path segments.

TRAVEL IMPACTS

		METHODOLOGY/ASSUMPTIONS:
Δ Vehicle trips:	-212 /day	Work trips = 3,929 workers in service area x 1.0% bicycle commuting mode share = 39 one-way trips. Non-work trips = 67 one-way trips
Δ VMT:	- 633 /day	
Δ Speed:	NA	Daily vehicle trips = (39 one-way work trips + 67 one-way non-work trips) x 2 = 212 daily trips.
Δ Delay:	NA	
Δ SOV	NA	Assume average trip is half the length of the bike facility.
Δ CP/VP	NA	Daily VMT reduction = (2 x 39 one-way trips) + (2 x 67 one-way trips) * (0.5 x 8.3 miles facility length) = 633 daily VMT reduction
Δ Transit	NA	
Δ Walk	NA	
Δ Bike	+ 1.0%	Work trips were calculated by estimating a 1 mile service area radius around the length of the 8.3 mile facility and then calculating the proportion of the total land of the community, the total population of the community, the number of households in the community, and the number of workers per household that would be served. The Bicycle Commuting Mode Share was estimated using the population density for the service area and a "Percent Bike Use for Commuting" table published by MassHighway Planning Department.

EMISSIONS

		METHODOLOGY/ASSUMPTIONS:
Δ VOC	- 0.5 kg/day	Emissions factor calculated from MOBILE5A, using 35 mph average commuter travel speed.
Δ NO$_x$	- 1.1 kg/day	
Δ CO	- 3.0 kg/day	VOC emissions factor = 0.819 g/mile
Δ PM$_{10}$	QA	NOx emissions factor = 1.672 g/mile
Δ PM$_{2.5}$	NA	Summer CO emissions factor = 5.096 g/mile
Δ Total	-1.6 kg/day	

COSTS

			Project life:__15__ yrs	Interest rate: __7__%

	CMAQ	NON-CMAQ	TOTAL	METHODOLOGY/ASSUMPTIONS:
Capital	$639,008	$660,902	$1,300,000	Assumes benefits 200 days/year.
Adm/oper	$0	$0	$0	
Total	$639,008	$660,902	$1,300,000	Cost-effectiveness was provided by the project sponsor, but calculated only on first year costs (not annualized).
Total annualized public cost:	$167,471			
Annual revenues:	None			
Net public cost:	$1,300,000			
Annual private cost	NA			
Total net cost	$1,300,000			

NOTE: A qualitative analysis for CO is listed for this project in the CMAQ database. However, a quantitative emissions reduction calculation for CO was provided by the State sponsor.

Category: **BICYCLE/PEDESTRIAN FACILITIES**		Subcategory: **Bicycle/Pedestrian**
CMAQ Project ID:IN20050009		Project Year: 2005
Location: Indiana		MPO: Michiana Area COG

Description: **4.3 Mile Bike Path to Pinhook Park** - Project constructed phase 1 of the Riverside Trail, a paved bike path from Angela Boulevard to Oakwood Boulevard. Phase 2 will begin early in 2008 and will complete the trail north to Darden Road where St. Joseph County is building a paved bike path to Indiana 933. When completed, the trail system will run 4.3 miles and allow cyclists and pedestrians located in the northwest portion of the city to access the downtown business district via connections with the East Race Walkway. It also provides pedestrians easier access to educational facilities at the University of Notre Dame and Indiana University South Bend using the River North Bikeway/Walkway.

TRAVEL IMPACTS

		METHODOLOGY/ASSUMPTIONS:
ΔVehicle trips:	- 83 /day (weekday, in-season)	Methodology based on guidance "Estimating the Effect of Bicycle Facilities on VMT and Emissions" prepared by the Seattle Engineering Department. Documentation notes uncertainty and difficulty in estimating new bicycle riders.
Δ VMT:	-249/day (weekday, in-season)	
ΔSpeed:	NA	Assume new bicycle riders equivalent to 1% of current drive alone workers in surrounding census tracks: 8,939 drive alone commuters x 1% x 2 trips each = 179 new bicycle trips diverted from driving.
ΔDelay:	NA	
ΔSOV	NA	
ΔCP/VP	NA	Assume that commuters divert 65% of the time: 179 bicyclists x .65 = 116 average new bike trips per day during season.
ΔTransit	NA	
ΔWalk	NA	Decreased autos = 116 new bike trips / 1.4 passengers per vehicle = 83 vehicle trips reduced per day during season.
ΔBike	+ 116 trip/day (weekday, in-season)	Assume 3 mile average trip length: Decreased VMT = 83 vehicle trips reduced x 3 mile average trip length = 249 VMT reduced per day during season.

Assumes seasonal use: 6 months per year. |

EMISSIONS

		METHODOLOGY/ASSUMPTIONS:
Δ VOC (HC)	- 0.37 kg/day	
Δ NO$_x$	- 0.45 kg/day	
Δ CO	- 2.65 kg/day	Emissions Factors from MACOG's Conformity Analysis from the Mobile 5A model with year 2000 socioeconomic data were used.
Δ PM$_{10}$	NA	
Δ PM$_{2.5}$	NA	
ΔTotal	- 0.82 kg/day (0.00090 tpd)	

COSTS

		Project life: __ 15 _ yrs		Interest rate: __ 7 __%

	CMAQ	NON-CMAQ	TOTAL	METHODOLOGY/ASSUMPTIONS:
Capital	$1,600,000	$ 400,000	$2,000,000	Estimated seasonal use is 132 days each year (6 months per year x 22 days per month). Multiple daily figures by 132 to get annual emissions effects.
Adm/oper	0	0	0	
Total	$1,600,000	$ 400,000	$2,000,000	
Total annualized public cost:	$237,332			The total project cost for phases I and II will be $3,500,000. CMAQ funding for Phase II will be $1,200,000 with an additional $300,000 local match.
Annual revenues:	None			
Net public cost:	$2,000,000			In order to calculate cost-effectiveness, assume the project life is 20 years.
Annual private cost	NA			
Total net cost	$2,000,000			

NOTE: Although project methodology and assumptions are well documented in an analysis report, the presentation of the calculation varies slightly from what is presented above. The calculation steps provided by the project sponsor were reordered to make the results clearer. In the reported evaluation, the results are presented in annual figures and order of steps differs.

Category: **BICYCLE/PEDESTRIAN FACILITIES**	Subcategory: **Bicycle Pedestrian**

CMAQ Project ID: Not Yet Assigned	Project Year: 2006
Location: Fort Collins, Colorado	MPO: North Front Range MPO

Description: **Construction of a Transit Bike Depot** - This project will fund the construction of a secure bicycle parking facility which will encourage the use of bicycling trips for work and to everyday destinations. The bike depot will have an attendant to oversee the area, and will also provide rentals, repairs, maintenance and safety information, restrooms and changing areas, bus pass sales, and other services geared toward bicyclists and transit users. The site may host community events and contain a small cafe or retail business in the future; the facility's location would provide easy access to other multi-modal connections.

TRAVEL IMPACTS

Travel Impact	Value	METHODOLOGY/ASSUMPTIONS:
Δ Vehicle trips:	- 120 /day	Assumes an increase of 160 average total daily trips by bicycle.
Δ VMT:	- 480 /day	Assumes the proportion of users that formerly commuted by SOV is 0.75.
Δ Speed:	NA	Assumes average one-way trip distance is 4 miles.
Δ Delay:	NA	
Δ SOV	NA	
Δ CP/VP	NA	Daily vehicle trip reduction = 160 x 0.75 = 120 vehicle trips reduced daily.
Δ Transit	NA	Daily VMT reduction = 160 added participants x 0.75 x 4 miles = 480 daily miles reduced.
Δ Walk	NA	
Δ Bike	NA	

EMISSIONS

Emission	Value	METHODOLOGY/ASSUMPTIONS:
Δ VOC	- 0.9072 kg/day	Average Fort Collins Network Vehicle speed is 25.4 mph.
Δ NO$_X$	- 0.9072 kg/day	Emissions reductions calculated by multiplying VMT reduction by vehicle emissions
Δ CO	- 6.6768 kg/day	VOC emissions reduction = 480 VMT x 0.00189 kg/mile = 0.9072 kg/day
Δ PM$_{10}$	NA	NOx emissions reduction = 480 VMT x 0.00189 kg/mile = 0.9072 kg/day
Δ PM$_{2.5}$	NA	CO emissions reduction = 480 VMT x 0.01391 kg/mile = 6.6768 kg/day
Δ Total	- 1.8144 kg/day (0.0020000 tpd)	

COSTS

			Project life:__6__ yrs	Interest rate: __7___%

	CMAQ	NON-CMAQ	TOTAL	METHODOLOGY/ASSUMPTIONS:
Capital	$ 63,910	$536,090	$600,000	Project assumes benefits 252 days per year and a total project life of 6 years.
Adm/oper	$0	$0	$0	
Total	$ 63,910	$536,090	$600,000	2 measures of cost-effectiveness were reported:
Total annualized public cost:	$131,797			Total Program Cost-Effectiveness (kg/$) CMAQ Cost-Effectiveness (kg/$)
Annual revenues:	None			
Net public cost:	$600,000			
Annual private cost	NA			
Total net cost	$600,000			

NOTE: The State provides a spreadsheet that automatically calculates emissions reductions, based on several set assumptions, as well as assumptions entered by the MPO.

Category: **BICYCLE PEDESTRIAN PROJECTS**	Subcategory: **Bicycle Pedestrian**

CMAQ Project ID: Not Yet Assigned	Project Year: 2007
Location: New York	MPO: New York Metropolitan Transportation Council

Description: **NYC CyclistNET Marketing Program** - This project will fund the creation of NYCyclistNet, a Web-based application consisting of a cycling parking locator, cycling tour maps, and an interactive routing system. NYCyclistNet will allow cyclists to create and plan bicycle routes, as well as to find up-to-date bicycle parking information and download a series of bicycle tours of the city from the Internet. The project will also provide City bike planners to collect and analyze information about the bicycle network and improve system performance.

TRAVEL IMPACTS

		METHODOLOGY/ASSUMPTIONS:
Δ Vehicle trips:	- 902 /day	Assume the average diverted bike trip is 4 miles and the number of daily bike users in NYC increases from 120,000 before the project to 121,200 after implementation.
Δ VMT:	- 3,608 /day	
Δ Speed:	NA	
Δ Delay:	NA	
Δ SOV	NA	Auto trip reduction = 1,200 new riders / 1.33 average vehicle occupancy = 902.
Δ CP/VP	NA	
Δ Transit	NA	Existing AADT for affected roadways in all five boroughs is 143,900. Calculate the decrease in AADT after implementation = 1,200 new bike users / 1.33 average vehicle occupancy / 0.627 short trip factor/ 0.01 diversion factor = 142,998.
Δ Walk	NA	
Δ Bike	NA	
		VMT reduced = 902 trips reduced * 4 mile trip = 3,608 miles.

EMISSIONS

		METHODOLOGY/ASSUMPTIONS:
Δ VOC	- 2.37 kg/day	Emissions reductions calculated using the CMAQtraq program developed by NYSDOT using the "BikePed Bikeway" module. Running emissions factors were used at 25 mph in each of the boroughs of New York City. Analysis assumes 190 work days per year.
Δ NO$_x$	- 1.96 kg/day	
Δ CO	- 38.43 kg/day	
Δ PM$_{10}$	- 0.9482 kg/day	
Δ PM$_{2.5}$	- 0.0426 kg/day	
Δ Total	- 4.33 kg/day	

COSTS

Project life:__10__ yrs   Interest rate: ___7__%

	CMAQ	NON-CMAQ	TOTAL	METHODOLOGY/ASSUMPTIONS:
Capital	$0	$0	$0	The project will be funded and implemented over four years (FY2007 – 2010). The total project life is 10 years and assumes benefits 190 work days per year.
Adm/oper	$2.4 M	$600,000	$3.0 M	
Total	$2.4 M	$600,000	$3.0 M	
Total annualized public cost:	$434,800			
Annual revenues:	None			
Net public cost:	$3.0 M			
Annual private cost	NA			
Total net cost	$3.0 M			

Category: **TRANSIT IMPROVEMENTS**		Subcategory: **New Bus Services**

CMAQ Project ID: WI20000004	Project Year: 2001
Location: Racine, Wisconsin	MPO: Milwaukee-Racine

Description: **City of Racine New Sunday Bus Service** - This project will expand the current bus service in the City of Racine by instituting Sunday service hours. It is expected that service would run from 8 AM to 4 PM, and would be provided over eight routes within the City of Racine on an hourly basis, using nine buses. Morning trips would focus on church-related activities and afternoon trips on shopping and social activities.

TRAVEL IMPACTS

		METHODOLOGY/ASSUMPTIONS:
Δ Vehicle trips:	- 72/day (Sunday only)	
Δ VMT:	NA	Assumes that 9 new bus trips per vehicle hour will be generated on the new service, and would replace drive alone trips.
Δ Speed:	NA	
Δ Delay:	NA	
Δ SOV	NA	Vehicle trip reduction = 9 new bus trips per vehicle hour x 8 hours of service = 72 trips reduced each Sunday.
Δ CP/VP	NA	
Δ Transit	+ 72/day (Sunday)	
Δ Walk	NA	
Δ Bike	NA	

EMISSIONS

		METHODOLOGY/ASSUMPTIONS:
Δ VOC	- 2.9 kg/day	
Δ NO$_X$	- 3.2 kg/day	Emissions reductions were calculated by multiplying VMT reduction by per-mile emissions factors for a typical summer day, based on MOBILE.
Δ CO	NA	
Δ PM$_{10}$	NA	
Δ PM$_{2.5}$	NA	
Δ Total	-6.1 kg/day	

COSTS

				Project life:__ 1 _ yrs	Interest rate: __ 7__%

	CMAQ	NON-CMAQ	TOTAL	METHODOLOGY/ASSUMPTIONS:
Capital	$0	$0	$0	Cost-effectiveness was not provided by sponsor. In order to calculate cost-effectiveness, assume the project has benefits 52 days per year, since emissions reductions were calculated for each Sunday.
Adm/oper	$157,382	$39,345	$196,727	
Total	$157,382	$39,345	$196,727	
Total annualized public cost:	$196,727			
Annual revenues:	NA			
Net public cost:	$196,727			
Annual private cost	NA			
Total net cost	$196,727			

NOTE: The project documentation could not provide the assumptions for average trip length and how they estimated 9 new bus trips per vehicle hour. The value seems low, given that the service will include 8 routes, implying only about one passenger per each bus. It should be noted that the daily emissions reduction is for each Sunday (emissions rates are based on summer weather conditions).

Category: **TRANSIT IMPROVMENTS**		Subcategory: **New Bus Service**
CMAQ Project ID: NY20050028	Project Year: 2005	
Location: Long Island, New York	MPO: New York Metropolitan Transportation Council	

Description: **Expanded S 92 Bus Route** - This project will fund upgrades to the current Monday – Saturday Suffolk County Transit (SCT) bus route S 92, improving bus frequencies during AM and PM peak periods from 1 hour to 30 minutes. Prior to July 2004, this bus route operated approximately 1,505 miles of daily revenue service between Orient Point/Greenport and East Hampton via Riverhead, Southampton Village, and Sag Harbor. The 70 mile long bus line regularly reaches capacity halfway along the route, leaving riders behind to catch taxis and carpool rides from other drivers. In some instances, employers from Southampton would travel to Riverhead and the North Fork to collect workers due to the uncertainty of the transit service. This project would provide additional AM and PM peak period bus trips at 30 minute frequencies – 4 AM trips and 4 PM trips in each direction. The added trips will increase passenger capacity and the 30 minute interval will offer more flexible travel times.

TRAVEL IMPACTS

		METHODOLOGY/ASSUMPTIONS:
Δ Vehicle trips:	- 358 /day	The additional ridership with this project is estimated at 200 passengers/day on weekdays and 160 on Saturdays. The estimated average passenger trip length is 19 miles. Travel speeds along the route are estimated as 40 mph.
Δ VMT:	- 6,640 /day	
Δ Speed:	NA	
Δ Delay:	NA	Autos daily vehicle trip change = - 200 weekday passengers + 160 weekend passengers = 360 trips
Δ SOV	NA	
Δ CP/VP	NA	Autos daily VMT change = 360 trips x 19 mile trip length = 6,840 VMT reduction.
Δ Transit	+ 2 bus trips/day	There will be 2 new roundtrip bus trips each day with the new service. The bus travels 140 miles on each roundtrip and averages 40 mph.
Δ Walk	NA	Bus daily trip change = + 2 bus trips/day
Δ Bike	NA	Bus daily VMT change = + 280 miles/day

Total daily vehicle trip change = 360 auto trips reduced – 2 new bus trips = 358 daily vehicle trip reduction.

Total daily VMT change = 6,849 auto VMT reduction – 280 bus VMT increase = 6,640 daily VMT reduction.

EMISSIONS

		METHODOLOGY/ASSUMPTIONS:
Δ VOC	- 6.66 kg/day	The NYSDOT software package CMAQtraq was used; the "Vehicle Reduction due to Transit" module to estimate emissions reductions from passenger vehicles and the "New Transit Bus" module to estimate emissions increases from the new bus service. Effects were calculated for 260 days/year with the following emissions factors (g/mile):
Δ NOx	+ 7.22 kg/day	
Δ CO	- 153.39 kg/day	
Δ PM10	+ 0.96 kg/day	
Δ PM2.5	+ 1.00 kg/day	
Δ Total	NA	

CO: Autos = 16.07   Bus = 6.30
VOC: Autos = 0.72   Bus = 0.98
NOx: Autos = 0.71   Bus = 17.13
PM2.5: Autos = 0.0133   Bus = 1.4543
PM10: Autos = 0.0269   Bus = 1.5926

Total Emissions = (Autos emissions factor * 988,000 Autos miles) - (Bus emissions factor * 72,800 bus miles) / 1,000

COSTS

	Project life:__3__ yrs	Interest rate: ___7__%

	CMAQ	NON-CMAQ	TOTAL	METHODOLOGY/ASSUMPTIONS:
Capital	$0	$0	$0	This project will have 260 days of operation. The project cost is for 3 years of operating subsidy. The non-CMAQ portion of the project cost will be funded by the local government with $90,000 in transit fares and county funding for $66,000.
Adm/oper	$264,000	$156,000	$420,000	
Total	$264,000	$156,000	$420,000	
Total annualized public cost:	$135,907			
Annual revenues:	$90,000			
Net public cost:	$330,000			
Annual private cost	NA			
Total net cost	$330,000			

NOTE: The $264,000 CMAQ funding amount provided by the project sponsor does not match the $336,000 listed in the CMAQ database. The emissions reductions estimates provided by the project sponsor also do not match those listed in the database (-2.5 kg/day VOC, +2.1 kg/day NOx, -59.3 kg/day CO, -0.3 kg/day PM_{10}, and -0.4 kg/day $PM_{2.5}$). The calculations in the document provided for this project did not account for the additional 160 weekend passengers; it only calculated Vehicle Trip Reduction (-200/day) and VMT Reduction (-3,800/day) for 200 additional passengers. Within the project documentation, the project is described as providing an additional four AM trips and four PM trips; however, the project calculations only account for 2 additional trips per day.

Category: **TRANSIT IMPROVEMENTS**			Subcategory: **New Bus Services**
CMAQ Project ID: RI20050003		Project Year: 2005	
Location: Rhode Island		MPO: Rhode Island State Planning Council	
Description: **Expanded Route 30 and New Route 12** - This program developed and implemented new transit operations expected to increase transit usage. A new express portion was added to Route 30, and a new Route 12 was created.			

Travel impacts (For New Express Portion of Route 30):			
Δ Vehicle trips:	- 53/day	Methodology/Assumptions (For Route 30):	
Δ VMT:	- 771/day	Calculate daily ridership increase due to new express portion of Route 30 service:	
Δ Speed:	NA	64 persons (1,049 with new service – 985 with old service).	
Δ Delay:	NA	Daily vehicle reduction = 64 new $\div$ 1.2 persons/vehicle = 53 daily vehicle	
Δ SOV	- 874 /day	reduction.	
Δ CP/VP	NA	Assumes average Route 30 round trip = 14.42 mi (calculated based on half of non-	
Δ Transit	+ 64 person/day	express portion of route + express portion of route). Daily VMT reduction = 53 vehicle trips reduced * 14.42 miles = 771 VMT reduction	
Δ Walk	NA	per day.	
Δ Bike	NA		

Emissions (For New Express Portion of Route 30):			
Δ VOC	- 0.92 kg/day	Methodology/Assumptions (For Route 30):	
Δ NO$_X$	- 1.05 kg/day		
Δ CO	- 17.89 kg/day	Emissions reductions were calculated by multiplying VMT reduction by per-mile	
Δ PM$_{10}$	NA	emissions factors.	
Δ PM$_{2.5}$	NA	Assumes speed = 35 mph.	
Δ Total	-2.0 kg/day	CO emissions calculated for winter.	

Travel impacts (For New Route 12)			
Δ Vehicle trips:	- 156 /day	Methodology/Assumptions (For Route 12):	
Δ VMT:	- 6,638 /day		
Δ Speed:	NA	CALCULATE TRAVEL ON OLD ROUTE: Utilization rates determined from existing	
Δ Delay:	NA	ridership. (448 persons). Daily vehicle reduction = 448 persons/day $\div$ 1.2	
Δ SOV	NA	persons/vehicle = 373 daily vehicle reduction.	
Δ CP/VP	NA	Roundtrip length of old Route 12 trip = 15.9 mi.	
Δ Transit	+ 187 /day	Daily VMT reduction = 373 daily vehicle reduction * 15.9 miles = 5,932 daily VMT	
Δ Walk	NA	reduction	
Δ Bike	NA	Calculate Travel on New Route: Total daily ridership for Expanded Route 12 = 635 persons/day. Daily vehicle reduction = 635 persons/day $\div$ 1.2 persons/vehicle = 529 daily vehicle reduction NEW Route 12 Roundtrip length = 23.74 mi. Daily VMT reduction = 529 vehicle reduction * 23.74 miles = 12,570 daily VMT reduction Difference equals net change due to new route.	

Emissions (For New Route 12):			
Δ VOC	- 5.78 kg/day	Methodology/Assumptions (For Route 12):	
Δ NO$_X$	- 10.00 kg/day	Emissions were calculated by multiplying VMT by per-mile emissions factors from	
Δ CO	- 173.11 kg/day	MOBILE in the base case (with old route) and with the expanded route.	

Δ PM$_{10}$	NA	Assumes: Speed = 35 mph for roadways where miles reduced under old route; Speed = 55 mph for roadways where miles reduced with expanded route.
Δ PM$_{2.5}$	NA	
Δ Total	NA	
		Emissions reductions with old route: 7.06 kg/day VOC, 137.69 kg/day CO (winter), 8.07 kg/day NOx. Emissions reductions with expanded route: 12.84 kg/day VOC, 310.79 kg/day CO (winter), 18.07 kg/day NOx. CO emissions only for winter.

Emissions (for Route 30 + Route 12)

Δ HC	NA	Methodology/Assumptions:
Δ VOC	- 6.7 kg/day	
Δ NO$_X$	- 11.1 kg/day	Sum totals from both routes
Δ CO	- 191.0 kg/day	
Δ PM$_{10}$	NA	
Δ PM$_{2.5}$	NA	
Δ Total	-17.8 kg/day	

Costs

| Project life:_2_ yrs | | | Interest rate: __ 7 __% |

	CMAQ	Non-CMAQ	Total	Methodology/Assumptions:
Capital	$440,000	$110,000	$550,000	
Adm/oper	$0	$0	$0	Annualized cost depends on whether costs include capital or only operating expenses.
Total	$440,000	$110,000	$550,000	
Total annualized public cost:	$328,780			
Annual revenues:	NA			
Net public cost:	$550,000			
Annual private cost	NA			
Total net cost	$550,000			

NOTE: Calculation methodology does not take into account changes in bus emissions associated with changes to the bus services.

Category: **TRANSIT IMPROVEMENTS**	Subcategory: **New Rail Services**

CMAQ Project ID: UT20020001	Project Year: 2002
Location: Utah	MPO: Wasatch Front Regional Council

Description: **Purchase of 5 New Light Rail Vehicles** - This project is the expansion of the current Light Rail service through the purchase of 5 new Light Rail Vehicles for its TRAX North/South line.

TRAVEL IMPACTS

		METHODOLOGY/ASSUMPTIONS:
Δ Vehicle trips:	- 2,046 /day	Assumes 1,023 daily new light rail round-trip riders x 2 trips per day = 2,046 vehicle trips reduced.
Δ VMT:	- 24,552 /day	
Δ Speed:	NA	Daily VMT reduction = 1023 new riders * 24 average round trip = 24,552 VMT reduced.
Δ Delay:	NA	
Δ SOV	NA	
Δ CP/VP	NA	
Δ Transit	NA	
Δ Walk	NA	
Δ Bike	NA	

EMISSIONS

		METHODOLOGY/ASSUMPTIONS:
Δ VOC	- 27.0 kg/day	Emissions reductions calculated by applying passenger car CO, NOx, and VOC g/mile rates for freeways and arterials.
Δ NO$_X$	- 33.0 kg/day	
Δ CO	- 305.0 kg/day	
Δ PM$_{10}$	NA	Congestion was measured by converting VMT reduction to Annual Vehicle Hours reduced: 75% on congested freeways @ 35 mph, 25% on congested arterials @ 17 mph.
Δ PM$_{2.5}$	NA	
Δ Total	-60.0 kg/day	

COSTS

	Project life:_20_ yrs		Interest rate: ___7__%

	CMAQ	NON-CMAQ	TOTAL	METHODOLOGY/ASSUMPTIONS:
Capital	$4,000,000	$0	$4,000,000	Cost benefit was calculated using a weighted ranking system that considered the following:
Adm/oper	$0	$0	$0	6) (10%) Project in a congested corridor
Total	$4,000,000	$0	$4,000,000	7) (15%) Length (years) of project effectiveness
Total annualized public cost:	$443,012			8) (25%) Emissions reduction
Annual revenues:	None			9) (25%) Congestion reduction (VHT)
Net public cost:	$4,000,000			10) (25%) Cost
Annual private cost	NA			This objective ranking was then combined with subjective rankings by staff and 3 different committees consisting of city planners and elected officials.
Total net cost	$4,000,000			

Category: **TRANSIT IMPROVMENTS**	Subcategory: **New Rail Services**
CMAQ Project ID: TX20030147	Project Year: 2003
Location: Dallas, Texas	MPO: North Central Texas Council of Governments

Description: **TRE Double Tracking of Segments** - This project will fund the construction of double tracking segments of the Trinity Railway Express (TRE) route between Dallas and Fort Worth. This project was selected during the 1999 Call for Projects held by the Dallas-Ft. Worth MPO and selected based on the demonstrated costs per ton of NOx emissions reduced. The project is part of a long-range plan by TRE to continue adding capacity through new siding construction. The strategy reduces emissions by providing new rail system services and/or expanding existing services to increase overall system ridership. The reduction methodology is adapted from "The Texas Guide to Accepted Mobile Source Emissions Reduction Strategies" published by Texas Transportation Institute, 2003.

TRAVEL IMPACTS

		METHODOLOGY/ASSUMPTIONS:
ΔVehicle trips:	- 5,400 /day	Assume 80% of new transit riders were previously vehicle drivers. Estimate new transit ridership will be 6,750 people. (6,750 riders * 0.8 = 5,400 vehicle trips reduced)
ΔVMT:	- 108,000/day	
ΔSpeed:	NA	
ΔDelay:	NA	Assume the average auto trip length is 20 miles (5,400 vehicle trips reduced x 20 miles = 108,000 vehicle miles reduced per day).
ΔSOV	NA	
ΔCP/VP	NA	
ΔTransit	NA	Assume each transit vehicle takes 5 trips per day and travels an average of 679 miles.
ΔWalk	NA	
ΔBike	NA	

EMISSIONS

		METHODOLOGY/ASSUMPTIONS:
Δ VOC	- 67.18 kg/day	Emissions reductions calculated using Mobile6 emissions factors, assuming a 34 mph running speed on all roadway types. Speed-based running exhaust emissions factor for roadways before project = NOx: 1.00 and VOC 0.56 grams/mile. Auto trip-end emissions factors = NOx: 0.39 and VOC: 1.25 grams/trip. Transit trip-end emissions factors for NOx and VOC are 0.0 grams/mile because starting emissions factors are not associated with the diesel locomotives that will be used in the TRE system equipment.
Δ NO$_X$	- 110 kg/day	
Δ CO	Kg	
Δ PM$_{10}$	NA	
Δ PM$_{2.5}$	NA	
Δ Total	NA	
		Reduction in daily auto start emissions from trips reduced = 5,400 trips x auto trip-end emissions factor for NOx and VOC. Reduction in daily auto running emissions = 108,000 miles x speed based running exhaust emissions factor for roadways for NOx and VOC. Increase in daily emissions from additional train starts = 5 transit trips * transit trip-end emissions factor (zero).

COSTS

Project life:__20__ yrs   Interest rate: ___7___%

	CMAQ	NON-CMAQ	TOTAL	METHODOLOGY/ASSUMPTIONS:
Capital	$36,253,821	$34,218,521	$70,472,342	11449 - TRE Quiet Zones and Quad Gates $4,174,000 total ($3,339,200 STP-MM, $834,800 local)
Adm/oper	$0	$0	$0	
Total	$36,253,821	$34,218,521	$70,472,342	

Total annualized public cost:	$7,631,000	TRE Elevated Double Tracked Section from West of Irby to Gilbert Rd: $66,298,342 total ($36,253,821 CMAQ, $1,328,585 STP-MM, $2,519,859 TxDOT PASS Funds, $26,196,077 local).
Annual revenues:	None	
Net public cost:	$70,472,342	
Annual private cost	NA	
Total net cost	$70,472,342	A cost effectiveness calculation was not provided by the project sponsor.

NOTE: Emissions reductions reported in the CMAQ database are different than those reported by the State because assumptions and emissions factors have been updated since the original calculations (-458 kg/day VOC, -78.1 kg/day NOx, -479.3 CO). The calculation does not account for any increase in emissions from the commuter rail.

Category: **TRANSIT IMPROVEMENTS**	Subcategory: **New Rail Services**

CMAQ Project ID: CT20050027	Project Year: 2005
Location: Fairfield, Connecticut	MPO: Greater Bridgeport Regional Planning Agency

Description: **Construct Rail Station Platforms and Bridge** - This project will fund the construction of a new commuter rail station, the Fairfield Metro-North Railroad station. The project will serve the residents of Fairfield, Connecticut, including students of Fairfield University — as well as nearby areas such as Black Rock within the city of Bridgeport — via the New Haven Line. The station will be a joint development, with a developer providing parking spaces and the State providing the railroad platform and an access roadway.

TRAVEL IMPACTS

		METHODOLOGY/ASSUMPTIONS:
Δ Vehicle trips:	0	Assumes 1,200 new parking spaces for rail patrons.
Δ VMT:	- 15,792 /day	Assumes 1/3 of the total ridership would be from new riders (based on rail ridership forecasts prepared by the Department for the new West Haven/Orange Rail Station Study).
Δ Speed:	NA	Roundtrip distances based on data from the Department's 2000 AM Peak Rail Survey. Of the Fairfield resident users, 21% destined to points within Connecticut, 79% destined to New York.
Δ Delay:	NA	
Δ SOV	NA	
Δ CP/VP	NA	
Δ Transit	NA	
Δ Walk	NA	Vehicle trip reduction =1,200 parking spaces x 1/3 new ridership utilization = 400 daily round trips reduced (no trip starts reduced).
Δ Bike	NA	VMT reduction = (400 vehicle trips reduced x 21% x 30 miles) + (400 vehicle trips reduced x 79% x 42 miles) = 2520 + 13,272 = 15,792 VMT reduced daily.

EMISSIONS

		METHODOLOGY/ASSUMPTIONS:
Δ VOC	- 6.0 kg/day	Emissions reductions calculated using Mobile6.2 with an average speed of 50 mph.
Δ NO$_X$	- 6.0 kg/day	
Δ CO	Kg	
Δ PM$_{10}$	NA	*Trips within Connecticut (30 miles roundtrip):*
Δ PM$_{2.5}$	- 1.0 kg/day	Daily emissions reduction = VMT x emissions factor.
Δ Total	- 12.0 kg/day (0.0132 tpd)	*New York-destined trips (42 miles roundtrip):* Daily emissions reduction = VMT x emissions factor.

COSTS

		Project life:__30__ yrs		Interest rate: __7___%

	CMAQ	NON-CMAQ	TOTAL	METHODOLOGY/ASSUMPTIONS:
Capital	$2.4 M	$600,000	$3.0 M	Assume benefits 260 days per year.
Adm/oper	$0	$0	$0	
Total	$2.4 M	$600,000	$3.0 M	Cost-effectiveness was not provided by the project sponsor.
Total annualized public cost:	$261,300			Annualized Cost = $3.0M x 0.081 CRF = $37,000 (assuming no private costs and no parking revenue). C/E = $37,000 / (0.0132 x 260) = $10,781/ton.
Annual revenues:	None			
Net public cost:	$3.0 M			
Annual private cost	NA			
Total net cost	$3.0 M			

NOTE: Emissions reductions reported in CMAQ database differ from estimates provided or calculated from sponsor-provided documentation (-12 kg/day VOC and -12 kg/day NOx). This project description does not distinguish between public and private costs and revenues, although there will presumably be cost-sharing and/or revenues (i.e. Parking fee revenues) between the public and private sectors.

Category: **TRANSIT IMPROVEMENTS**	Subcategory: **Service Upgrades/Amenities**
CMAQ Project ID: MA20020069	Project Year: 2002
Location: Fitchburg, Massachusetts	MPO: Montachusett Regional Planning Commission

Description: **Fitchburg ITC Parking Garage** - This project is the construction of a multi-level bus circulation/parking garage adjacent to the existing MART Intermodal Transportation Center on Main Street in Fitchburg. The project will provide parking spaces for about 387 cars, intra- and inter-city bus circulation system, fare collection system, and interface with the regional bus service and commuter rail station. The project will also provide an additional 13,000 square feet of retail space in the new facility to be rented out to commercial establishments, including banking, restaurants, and dry cleaners.

TRAVEL IMPACTS

		METHODOLOGY/ASSUMPTIONS:
Δ Vehicle trips:	0	387 total parking spaces – 25 spaces for MART staff – 35 spaces for commercial development = 327 new spaces for commuter rail passengers. Assume 100% utilization rate of all 327 new spaces, because demand at the Fitchburg Commuter Rail Station will increase 6% per year and eventually reach an estimated 410 one-way rail passengers.

Using a 1999 survey of rail riders in the Montachusett area, 89.5% of passengers travel to the Cambridge/Boston area, a one-way trip of 43 miles. Based on the survey, conservatively assume that 75% of the 327 new vehicles parked at the facility will have as their final destination Cambridge/Boston. 327 vehicles * 75% = 245 vehicle round trips removed = 490 vehicle trips removed (no trip starts reduced).

Daily VMT reduction = 245 vehicle round trips removed * 86 mile round trip = 21,070 VMT reduced. |
Δ VMT:	- 21,070 /day	
Δ Speed:	NA	
Δ Delay:	NA	
Δ SOV	NA	
Δ CP/VP	NA	
Δ Transit	NA	
Δ Walk	NA	
Δ Bike	NA	

EMISSIONS

		METHODOLOGY/ASSUMPTIONS:
Δ VOC	- 14.0 kg/day	Emissions reductions calculated using Mobile 5a emissions factors. Assumes average travel speed along the Route 2 corridor is 50 mph.

Note: The calculated CO emissions reduction is only for winter months. |
Δ NO_x	- 27.0 kg/day	
Δ CO	- 143.0 kg/day	
Δ PM_{10}	NA	
Δ $PM_{2.5}$	NA	
Δ Total	-41 kg/day (0.0451 tpd)	

COSTS

Project life:____ yrs	Interest rate: __7___ %

	CMAQ	NON-CMAQ	TOTAL	METHODOLOGY/ASSUMPTIONS:
Capital	$388,000	$237,000	$625,000	Assumes benefits 250 days/year.

The following cost-effectiveness was provided by the project sponsor. First year cost per kg of emissions reduced = Project cost / Adjusted net change (kg/year).

VOC cost effectiveness = $625,000 / 3,483 kg/year = $179
NOx cost effectiveness = $625,000 / 6,805 kg/year = $91
Winter CO cost effectiveness = $625,000 / 35, 197 kg/year = $18

Annualized cost = $625,000 x 0.081 CRF = $50,625
C/E = $50,625 / (0.0451 x 250) = $4,490/ton |
Adm/oper	$0	$0	$0	
Total	$388,000	$237,000	$625,000	
Total annualized public cost:	$625,000			
Annual revenues:	None			
Net public cost:	$625,000			
Annual private cost	NA			
Total net cost	$625,000			

NOTE: The emissions reduction methodology for the Intermodal Transportation Center does not account for the additional revenues from renting commercial space on the facility or the additional operating cost of the new facility.

Category: **TRANSIT IMPROVEMENTS**	Subcategory: **Service Upgrades/Amenities**
CMAQ Project ID: MO20040023	Project Year: 2004 - 2006
Location: Kansas City, Missouri	MPO: Mid-American Regional Council

Description: **Operation Welcome Aboard Infrastructure** - Operation Welcome Aboard is a program designed to increase transit ridership by improving the comfort, attractiveness & usefulness of bus shelters. The project will install 100 bus shelters and pads featuring the new paint scheme of Metro buses. Signage at stops will also be improved to tie in with color and provide valuable route and schedule information. By making bus stops more inviting and useful, new riders will be more likely to find out about transit services and use them.

Travel impacts

		Methodology/Assumptions:
△ Vehicle trips:	- 405/day	Assume New Bus Ridership will be 450 people/day and the average Bus Trip Length is 10 miles. Estimate that 90 percent of the new bus riders will be switching from autos. Estimates were derived from internal analysis.
△ VMT:	- 4,050/day	
△ Speed:	NA	
△ Delay:	NA	
△ SOV	NA	Vehicle trips reduced = 450 people/day * 0.9 new ridership factor = 405 vehicle trips/day
△ CP/VP	NA	VMT reduced = 405 vehicle trips x 10 miles = 4,050 VMT reduction
△ Transit	+ 450/day	
△ Walk	NA	Methodology based on December 1995 guidance from CARB entitled "Emissions Reduction Calculation Methodologies."
△ Bike	NA	

Emissions

		Methodology/Assumptions:
△ HC	NA	Emissions reductions were calculated by multiplying VMT reduction by per-mile emissions factors. Emissions factors developed for the Kansas City region based on MOBILE model; assumed running speed is 35 mph.
△ VOC	- 2.49 kg/day	
△ NO$_x$	- 3.38 kg/day	
△ CO	NA	
△ PM$_{10}$	NA	
△ PM$_{2.5}$	NA	
△ Total	NA	

Costs

Project life: _10_ yrs | Interest rate: _7_%

	CMAQ	Non-CMAQ	Total	Methodology/Assumptions:
Capital	$960,000	$240,000	$1.2 M	The multi-year project has joint sponsorship by Kansas and Missouri. Kansas will provide 10% of the CMAQ and local match, while Missouri assumes 90% of the CMAQ and local match funding.
Adm/oper	$0	$0	$0	
Total	$960,000	$240,000	$1.2 M	
Total annualized public cost:	$190,900			
Annual revenues:	None			
Net public cost:	$1.2 M			
Annual private cost	NA			
Total net cost	$1.2 M			

2004   *2005*   *2006*
KS20040011 $60   KS20040011 $20   KS20040011 $16
MO20040023 $540   MO20050009 $180   MO2006006 $144

To calculate cost effectiveness, need to assume life of project (recommend 10 years).

Category: **TRANSIT IMPROVEMENTS**	Subcategory: **Service Upgrades/Amenities**

CMAQ Project ID: NY20040006	Project Year: 2004
Location: Suffolk County, New York	MPO: New York Metropolitan Transportation Council

Description: **Suffolk County Transit Marketing Program** - The project will fund a suite of service upgrades and transit marketing tasks to inform the public of available transit options. These tasks will entail (1) review current marketing materials (2) schedule ads in local radio and newspaper outlets (3) inform the public of the new SCT color scheme (4) develop additional paratransit marketing materials and (5) develop and administer a rider survey.

TRAVEL IMPACTS

		METHODOLOGY/ASSUMPTIONS:
Δ Vehicle trips:	- 176 /day	
Δ VMT:	- 2,499.20 /day	Through these efforts SCT estimates the system will attract 176 new riders each day. The average trip length per rider is 14.2 miles per day and the average travel speed on local network is 18 mph. The system operates 307 days per year.

Daily Trip Reduction = - 176 trips/day x 14.2 miles = 2,499.2 VMT reduced. |
Δ Speed:	NA	
Δ Delay:	NA	
Δ SOV	NA	
Δ CP/VP	NA	
Δ Transit	NA	
Δ Walk	NA	
Δ Bike	NA	

EMISSIONS

		METHODOLOGY/ASSUMPTIONS:
Δ VOC	- 2.39 kg/day	The NYSDOT software package CMAQtraq was used to estimate emissions for the project, using the "Transit Vehicle Reduction" module at 18 mph running speeds. Effects were calculated for 307 days/year with the following emissions factors (g/mile):
CO: 16.29
VOC: 0.96
NOx: 0.87
PM2.5: 0.0133
PM10: 0.0269
Δ NOₓ	- 2.19 kg/day	
Δ CO	- 40.72 kg/day	
Δ PM$_{10}$	- 0.067 kg/day	
Δ PM$_{2.5}$	- 0.033 kg/day	
Δ Total	- 4.58 kg/day (0.0050 tpd)	

COSTS

Project life:_2__ yrs   Interest rate: ___7__%

	CMAQ	NON-CMAQ	TOTAL	METHODOLOGY/ASSUMPTIONS:
Capital	$0	$0	$0	
Adm/oper	$160,000	$40,000	$200,000	
Total	$160,000	$40,000	$200,000	
Total annualized public cost:	$123,600			
Annual revenues:	None			
Net public cost:	$200,000			
Annual private cost	NA			
Total net cost	$200,000			

Category: **TRANSIT IMPROVEMENTS**	Subcategory: **Service Upgrades/Amenities**
CMAQ Project ID: OH20050008	Project Year: 2005
Location: Cuyahoga County, Ohio	MPO: Northeast Ohio Areawide Coordinating Agency

Description: **Laketran AVL-MDT System** - Installation of automatic vehicle location (AVL) and mobile data terminal (MDT) systems on Laketran vehicles will improve transit vehicle operations as part of the system's paratransit program. These benefits may include improving schedule adherence, reducing operations cost, improving efficiency, increasing ridership, reducing the number of vehicles needed, and improving routes planning and scheduling.

TRAVEL IMPACTS

		METHODOLOGY/ASSUMPTIONS:
Δ Vehicle trips:	NA	Assumes a 7% reduction in the number of vehicles required to serve the same routes and passengers and a 15-18% decrease in travel time on transit.
Δ VMT:	-465,553 /year	
Δ Speed:	NA	Assume a 17.5% increase in Paratransit ridership (54,387 passengers) traveling an average of 8.56 miles per passenger. (465,553 VMT reduced)
Δ Delay:	NA	
Δ SOV	NA	
Δ CP/VP	NA	Annual VMT reduction = 54,387 passengers * 8.56 miles of service provided = 465,553 VMT reduction
Δ Transit	NA	
Δ Walk	NA	
Δ Bike	NA	

EMISSIONS

		METHODOLOGY/ASSUMPTIONS:
Δ VOC	- 4.0 kg/day	Emissions reductions were calculated using VMT reductions and EPA Standards in g/mile for HC, CO, NOx and PM in 2004.
Δ NO_x	- 13.0 kg/day	
Δ CO	- 47.0 kg/day	
Δ PM_{10}	NA	
Δ $PM_{2.5}$	NA	
Δ Total	-17.0 kg/day (0.0187 tpd)	

COSTS

Project life:__10__ yrs   Interest rate: __7__ %

	CMAQ	NON-CMAQ	TOTAL	METHODOLOGY/ASSUMPTIONS:
Capital	$2,800,000	$700,000	$3,500,000	The project is split between two years, $1.1 M in funding will be used in 2004 and $2.4 M in 2005. Cost-effectiveness was not provided by sponsor.
Adm/oper	$0	$0	$0	
Total	$2,800,000	$700,000	$3,500,000	
Total annualized public cost:	$538,580			Assume technology lasts 10 years.
Annual revenues:	None			
Net public cost:	$3,500,000			
Annual private cost	NA			
Total net cost	$3,500,000			

Category: **TRANSIT IMPROVEMENTS**	Subcategory: **Service Upgrades/Amenities**

CMAQ Project ID: Not Yet Assigned	Project Year: 2007
Location: Connecticut	MPO: South Central Regional COG

Description: Commuter Rail Utility Construction - This project will fund the construction of an additional 199 parking spaces at the Guilford Station-Woodruff Farms station along the Shoreline East Line. An AM Peak Rail Origin/Destination Survey was conducted in 2000 to determine the destination towns for and percentage of trips made by patrons boarding at the Guilford rail station.

TRAVEL IMPACTS

		METHODOLOGY/ASSUMPTIONS:
Δ Vehicle trips:	- 298 /day	VMT reductions were calculated using the length of highways in each county for each trip destination. Fairfield County had 30% of the VMT reduction (3,172 miles) and New Haven County had 70% of the VMT reduction (7,530 miles).
Δ VMT:	- 10,702 /year	
Δ Speed:	NA	Assume a 100% utilization of the proposed 199 parking spaces in Year 2007. Vehicle trip reduction = 199 spaces x 100% utilization x 2 trips per day = 298 vehicle trips reduced daily.
Δ Delay:	NA	
Δ SOV	NA	
Δ CP/VP	NA	
Δ Transit	NA	
Δ Walk	NA	
Δ Bike	NA	

EMISSIONS

		METHODOLOGY/ASSUMPTIONS:
Δ VOC	- 6.0 kg/day	Emissions factors from MOBILE model with 50 mph traveling speed.
Δ NO$_x$	- 6.0 kg/day	
Δ CO	Kg	*Fairfield County:*
Δ PM$_{10}$	NA	VOC emissions factor = 0.544 g/mile
Δ PM$_{2.5}$	- 1.0 kg/day	NOx emissions factor = 0.508 g/mile
Δ Total	-12.0 kg/day (0.0132 tpd)	PM2.5 emissions factor = 0.011 g/mile
		New Haven County:
		VOC emissions factor = 0.546 g/mile
		NOx emissions factor = 0.524 g/mile
		PM2.5 emissions factor = 0.011 g/mile

COSTS

Project life:__12__ yrs   Interest rate: ___7__%

	CMAQ	NON-CMAQ	TOTAL	METHODOLOGY/ASSUMPTIONS:
Capital	$89,000	$22,000	$111,000	Assumes project benefits 260 days each year.
Adm/oper	$0	$0	$0	
Total	$89,000	$22,000	$111,000	Cost-effectiveness was not provided by the project sponsor.
Total annualized public cost:	$14,227			
Annual revenues:	None			
Net public cost:	$111,000			
Annual private cost	NA			
Total net cost	$111,000			

NOTE: Project cost information taken from 2007 STIP.

Category: **TECHNOLOGY IMPROVMENTS**	Subcategory: **Conventional Bus Replacements**
CMAQ Project ID: MD20020008	Project Year: 2002
Location: Maryland	MPO: Metropolitan Washington COG

Description: **100 Replacement Local Buses** - Purchase of 100 conventionally fueled local buses for the Maryland Transit Administration (MTA) fleet. The new buses will replace 100 buses older buses that have been in operation since 1988.

Travel impacts

		Methodology/Assumptions:
Δ Vehicle trips:	0	
Δ VMT:	0	MTA buses operate an average of 330 days and travel an average of 26,650 miles each year. (26,650 miles * 100 buses = 2,665,000 VMT)
Δ Speed:	0	
Δ Delay:	0	
Δ SOV	0	
Δ CP/VP	0	Because the buses will replace existing buses, there are no estimated travel impacts.
Δ Transit	0	
Δ Walk	0	
Δ Bike	0	

Emissions

		Methodology/Assumptions:
Δ VOC	- 17.0 kg/day	Emissions savings were estimated using the difference between the emissions factors for the 1988 buses and the 2002 replacement buses:
Δ NO_X	- 188.9 kg/day	
Δ CO	NA	VOC: 4.660 g/mi (1988 bus) – 2.560 g/mi (2002 bus) = 2.100 g/mi savings
Δ PM_{10}	NA	NOx: 36.24 g/mi (1988 bus) – 12.88 g/mi (2002 bus) = 23.36 g/mi savings
Δ $PM_{2.5}$	NA	
Δ Total	NA	

VOC reduced = (2,665,000 VMT * 2.1 g/mi savings) / 330 days = 17.0 kg/day
NOX reduced = (2,665,000 VMT * 23.36 g/mi savings) / 330 days = 188.9 kg/day

Costs

Project life:__ 4 _ yrs   Interest rate: __ 7 __%

	CMAQ	Non-CMAQ	Total	Methodology/Assumptions:
Capital	$5.0 M	$21.5 M	$26.5 M	
Adm/oper	$0	$0	$0	Cost-effectiveness was not provided by sponsor. In order to calculate cost-effectiveness, estimate the remaining useful life of the older vehicles.
Total	$5.0 M	$21.5 M	$26.5 M	
Total annualized public cost:	$9,180,000			
Annual revenues:	None			
Net public cost:	$26,500,000			
Annual private cost	NA			
Total net cost	$26,500,000			

NOTE: Emissions reductions that come from replacing an older vehicle with a newer, cleaner vehicle will not provide emissions reduction credit longer than the period of time that the older vehicle would have been kept in service without the replacement program (per EPA's Diesel Retrofit SIP and Conformity guidance, http://www.epa.gov/cleandiesel/publications.htm.) The duration of benefit will therefore depend on the remaining life of the vehicles. This causes some difficulty in calculating cost-effectiveness, since the buses will be used for perhaps another 12+ years (and will allow for continued transit services), but the emissions benefits associated with replacement may only last for a couple years.

Strategy: **TECHNOLOGY IMPROVMENTS**	Category: **Conventional Bus Replacements**
CMAQ Project ID: Not Yet Assigned	Project Year: 2003
Location: Southwest Ohio	MPO: Ohio-Kentucky-Indiana Regional COG

Description: **61 Replacement Buses** - This project will fund the purchase of 61 new 40-foot coaches to replace 15-year old ones. The new coaches will reduce air pollution because they are manufactured to adhere to much stricter air quality standards than the coaches they replace. The coaches will be equipped with security cameras and bike racks to increase security and provide multimodal connectability. The coaches are lift-equipped for disability accessibility. They also come equipped with ITS equipment and METRO, which are connected to ARTIMIS, allowing the transfer of information on highways to aid in congestion relief.

TRAVEL IMPACTS

		METHODOLOGY/ASSUMPTIONS:
Δ Vehicle trips:	NA	Methodology does not account for any reduction in person motor vehicle travel, simply the replacement of existing buses. The methodology actually assumes an increase in VMT from the buses, as the new buses travel more.
Δ VMT:	+ 45 bus miles/day	
Δ Speed:	NA	
Δ Delay:	NA	Average daily VMT for old buses = 77 VMT is the default value for 15-year old urban transit buses using MOBILE 6.2.
Δ SOV	NA	
Δ CP/VP	NA	
Δ Transit	NA	Average daily VMT for new buses = 122 VMT is the default value for 1-year old urban transit buses using MOBILE 6.2.
Δ Walk	NA	
Δ Bike	NA	

EMISSIONS

		METHODOLOGY/ASSUMPTIONS:
Δ VOC	-9.639 kg/day	Calculation used MOBILE 6.2 emissions factors for 15-year old and 1-year old urban transit buses operating on local streets.
Δ NO$_x$	-11.639 kg/day	Emissions factors for 15-year old urban transit buses:
Δ CO	-35.530 kg/day	
Δ PM$_{10}$	NA	VOC = 2.74 g/mile; NOx = 24.20 g/mile; CO = 12.61 g/mile
Δ PM$_{2.5}$	NA	Emissions factors for 1-year old urban transit buses
Δ Total	NA	VOC = 0.44 g/mile; NOx = 10.59 g/mile; CO = 6.44 g/mile

Bus emissions are calculated by multiplying VMT by emissions factor.
Total old bus emissions – Total new bus emissions = Total emissions reduction
VOC: 12934 – 3295 = 9.639 kg/day
CO: 59499 – 47860 = 11.639 kg/day
NOX: 114234 – 78704 = 35.530 kg/day

COSTS

| | | | Project life: __4__ yrs | Interest rate: __7__ % |

	CMAQ	NON-CMAQ	TOTAL	METHODOLOGY/ASSUMPTIONS:
Capital	$4,864,440	$2,084,760	$6,949,200	A cost effectiveness calculation was not provided by the project sponsor. Emissions reductions that come from replacing an older vehicle with a newer, cleaner vehicle will not provide emissions reduction credit longer than the period of time that the older vehicle would have been kept in service without the replacement program (per EPA's Diesel Retrofit SIP and Conformity guidance.) The duration of benefit will therefore depend on the remaining life of the vehicles. This causes some difficulty in calculating cost-effectiveness, since the buses will be used for perhaps another 12+ years (and will allow for continued transit services).
Adm/oper	$0	$0	$0	
Total	$4,864,440	$2,084,760	$6,949,200	
Total annualized public cost:	$2,353,000			
Annual revenues:	None			
Net public cost:	$6,949,200			
Annual private cost	NA			
Total net cost	$6,949,200			

NOTE: The methodology uses MOBILE6.2 defaults, but do not appear to be based on actual travel data for these buses. It is unclear why the new buses would travel more than the old buses, so this seems to be a very conservative assumption (reduces the amount of emissions benefit). Emissions reductions reported in the CMAQ database differ from estimates provided or calculated from sponsor-provided documentation (-1 kg/day VOC, -60 kg/day CO, and -22 kg/day NOx).

Category: **TECHNOLOGY IMPROVEMENTS**	Subcategory: **Alternative Fuel Vehicles/Fueling Facilities**
CMAQ Project ID: ME20020020	Project Year: 2002
Location: Cumberland County, Maine	MPO: Portland Area Comprehensive Transportation Study

Description: **Compressed Natural Gas Fueling Station** - This project will fund the construction of a fast fill compressed natural gas facility for public and private fleets based in, or operating from the Greater Portland area. This project will be implemented in conjunction with an effort by METRO to convert the transit bus fleet to natural gas. The agency anticipates converting 4 buses by 2004 and a total of 21 buses by 2015.

TRAVEL IMPACTS

		METHODOLOGY/ASSUMPTIONS:
Δ Vehicle trips:	NA	METRO's existing buses use an average of 32 gallons of diesel fuel per day.
Δ VMT:	NA	
Δ Speed:	NA	
Δ Delay:	NA	2004: 4 buses * 32 gallons of fuel = 128 gallons/day
Δ SOV	NA	2015: 21 buses * 32 gallons of fuel = 672 gallons/day
Δ CP/VP	NA	
Δ Transit	NA	
Δ Walk	NA	
Δ Bike	NA	

EMISSIONS

		METHODOLOGY/ASSUMPTIONS:
Δ VOC	- 2.768 kg/day	Assumes that diesel engines emit 27.04 g/gal of VOC and 83.2 g/gal of NOx.
Δ NO$_X$	- 2.13 kg/day	Assumes that natural gas reduces VOC emissions by 80% over diesel fuel
Δ CO	NA	Assumes that natural gas reduces NOx emissions by 20% over diesel fuel
Δ PM$_{10}$	NA	
Δ PM$_{2.5}$	NA	
Δ Total	-4.90 kg/day (0.0054 tpd)	2004 VOC: 128 gallons/day * 27.04 g/gal * 0.8 = 2.768 kg/day
		2004 NOx: 128 gallons/day * 83.2 g/gal * 0.2 = 2.13 kg/day
		2015 VOC: 672 gallons/day * 27.04 g/gal * 0.8 = 14.536 kg/day
		2015 NOx: 672 gallons/day * 83.2 g/gal * 0.2 = 11.182 kg/day

COSTS

Project life:__12__ yrs   Interest rate: __7___%

	CMAQ	NON-CMAQ	TOTAL	METHODOLOGY/ASSUMPTIONS:
Capital	$150,000	$1,155,903	$1,305,903	
Adm/oper	$0	$0	$0	
Total	$150,000	$1,155,903	$1,305,903	
Total annualized public cost:	$192,912			
Annual revenues:	None			
Net public cost:	$1,305,903			
Annual private cost	NA			
Total net cost	$1,305,903			

Category: **TECHNOLOGY IMPROVEMENTS**	Subcategory: **Alternative Vehicles/ Fueling Facilities**
CMAQ Project ID: PA20020062	Project Year: 2002
Location: Philadelphia, Pennsylvania	MPO: Delaware Valley Regional Planning Commission

Description: **Purchase 12 New Alternative Fuel Bus** - This project will fund the acquisition of 12 forty-foot, low floor hybrid/electric-powered buses with an option to purchase 20 additional buses. These buses, through the combination of an internal-combustion engine to produce electricity, storage batteries, and an electric propulsion system, will provide a quieter ride for riders, reduce exhaust emissions and fuel consumption, and improve brake life through regenerative braking.

TRAVEL IMPACTS

		METHODOLOGY/ASSUMPTIONS:
Δ Vehicle trips:	NA	The 12 buses being retired used diesel fuel and had an annual 27,207 vehicle revenue miles per bus.
Δ VMT:	NA	
Δ Speed:	NA	
Δ Delay:	NA	The 12 replacement buses are hybrid/electric vehicles and will still have an annual 27,207 vehicle revenue miles per bus.
Δ SOV	NA	
Δ CP/VP	NA	
Δ Transit	NA	Total revenue miles are multiplied by a 1.15 deadhead factor to account for vehicle travel to and from the programmed bus routes.
Δ Walk	NA	
Δ Bike	NA	

EMISSIONS

		METHODOLOGY/ASSUMPTIONS:
Δ VOC	- 3.0 kg/day	Emissions factors were calculated from the Mobile model using an average running speed of 13.5 mph for both conventional and replacement buses.
Δ NO$_x$	- 91.0 kg/day	
Δ CO	- 12.0 kg/day	
Δ PM$_{10}$	NA	Emissions reductions were calculated for the year 2002, assuming operation 250 days per year using an interim version of PAQONE created specifically for DVRPC to calculate air quality benefits. A diesel fuel type was selected for the older buses and a CNG fuel type for the newer ones.
Δ PM$_{2.5}$	NA	
Δ Total	-94 kg/day (0.1035 tpd)	

COSTS

			Project life: __4__ yrs	Interest rate: __7__ %

	CMAQ	NON-CMAQ	TOTAL	METHODOLOGY/ASSUMPTIONS:
Capital	$5.608 M	$1.402 M	$7.010 M	Cost-effectiveness was not provided by sponsor.
Adm/oper	$0	$0	$0	
Total	$5.608 M	$1.402 M	$7.010 M	
Total annualized public cost:	$2,428,000			
Annual revenues:	None			
Net public cost:	$7,010,000			
Annual private cost	NA			
Total net cost	$7,010,000			

NOTE: The calculations provided in the documentation are unclear and might suggest that the formula for vehicle travel to/from bus routes = 108.8 daily mileage per bus x 12 buses x difference in emissions rates x 1.15 deadhead factor. Emissions reductions that come from replacing an older vehicle with a newer, cleaner vehicle will not provide emissions reduction credit longer than the period of time that the older vehicle would have been kept in service without the replacement program (per EPA's Diesel Retrofit SIP and Conformity guidance). The duration of benefit will therefore depend on the remaining life of the vehicles.

Strategy: **TECHNOLOGY IMPROVEMENTS**		Category: **Alternative Vehicles/ Fueling Facilities**
CMAQ Project ID: CT20050025		Project Year: 2005
Location: Connecticut		MPO: No MPO Identified/State-sponsored project

Description: CT Clean Fuels Program - This project is the funding of provision of technical assistance to municipalities and other entities that implement alternate fuel projects in the NY/NJ/CT non-attainment area – these include purchases of diesel particulate filters for buses and other equipment. The purchase and/or conversion of alternate fuel vehicles in the public or private sector are aimed primarily at air quality improvement. Alternate fuel vehicles replace conventionally powered vehicles, resulting in lower levels of controlled emissions.

TRAVEL IMPACTS

		METHODOLOGY/ASSUMPTIONS:
Δ Vehicle trips:	NA	Type of equipment = Diesel Particulate Filters.
Δ VMT:	NA	# of DPFs installed = 9, with plans to install 31 more.
Δ Speed:	NA	Average daily distance driven based on MOBILE6.2 estimates:
Δ Delay:	NA	Light duty trucks/vans = 34.3 miles/day
Δ SOV	NA	Passenger cars = 28.8 miles/day
Δ CP/VP	NA	Average number of days/week vehicle is used = 5 days/week.
Δ Transit	NA	
Δ Walk	NA	
Δ Bike	NA	

EMISSIONS

		METHODOLOGY/ASSUMPTIONS:
Δ VOC	- 6.75 kg/day	Emissions reduction benefits calculated using the Department of Energy's "AirCRED" model, with variables such as the number of vehicles, the type of vehicles, an estimate of average daily distance driven, and the average number of days per week the vehicle is used, entered into the model.
Δ NO$_X$	-12.49 kg/day	
Δ CO	NA	
Δ PM$_{10}$	NA	
Δ PM$_{2.5}$	NA	
Δ Total	-19.2 kg/day	NY/NJ/CT Moderate Ozone Non-Attainment Area: VOC emissions reduction = 0.10 lbs NMHC/day x 0.4536 lb/kg x 153 days = 6.94 kg/day. Then, convert NMHC to VOC = 6.94 kg/day / 0.93 x 0.45 = 3.36 kg/day. NOx emissions reduction = 0.06 lbs/day x 0.4536 lb/kg x 153 days = 4.16 kg/day. Greater Connecticut Moderate Ozone Non-Attainment Area: VOC emissions reduction = 0.11 lbs MHC/day x 0.4536 lb/kg x 153 days = 7.63 kg/day. Then, convert NMHC to VOC = 7.63 / 0.93 x 0.45 = 3.39 kg/day NOx emissions reduction = 0.12 lbs/day x 0.4536 lb/kg x 153 days = 8.33 kg/day

COSTS

				Project life: __7__ yrs	Interest rate: ___7__%

	CMAQ	NON-CMAQ	TOTAL	METHODOLOGY/ASSUMPTIONS: Assumes benefits for 153 days/year.
Capital	$688,800	$172,200	$861,000	
Adm/oper	$0	$0	$0	A cost effectiveness calculation was not provided by the project sponsor.
Total	$688,800	$172,200	$861,000	
Total annualized public cost:	$172,670			
Annual revenues:	None			
Net public cost:	$861,000			
Annual private cost	NA			
Total net cost	$861,000			

NOTE: Emissions reductions reported in CMAQ database (-11.4 kg/day VOC, -40.3 kg/day NOx) differ from estimates provided or calculated from sponsor-provided documentation.

Category: **TECHNOLOGY IMPROVEMENTS**	Subcategory: **Alternative Fuel Vehicles/ Fueling Facilities**
CMAQ Project ID: Not Yet Assigned	Project Year: 2007
Location: Nassau County, New York	MPO: New York Metropolitan Transportation Council
Description: **Purchase 3 Forty-Foot Urban Transit CNG Buses** - This project will fund the purchase of three replacement, 40-foot CNG transit buses. The present buses will be at the end of their useful life and continued use would result in increased maintenance, increased costs, and poor, inefficient service.	

TRAVEL IMPACTS

		METHODOLOGY/ASSUMPTIONS:
Δ Vehicle trips:	NA	
Δ VMT:	NA	The three buses being replaced are 1997 CNG buses which average 14 mph. Each bus travels 160 miles each day and operates 360 days per year.
Δ Speed:	NA	
Δ Delay:	NA	
Δ SOV	NA	
Δ CP/VP	NA	
Δ Transit	NA	
Δ Walk	NA	
Δ Bike	NA	

EMISSIONS

		METHODOLOGY/ASSUMPTIONS:
Δ VOC	- 1.50 kg/day	"After" analysis conducted using CO, VOC, and PM emissions factors from a 2004 model year 6081H John Deere engine test conducted by University of West Virginia for Washington Metropolitan Area Transit Authority (WMATA).
Δ NO$_x$	- 4.34 kg/day	
Δ CO	- 7.62 kg/day	
Δ PM$_{10}$	NA	
Δ PM$_{2.5}$	- 1.40 kg/day	"Before" emissions factors for CO, NOx, and VOC derived from 1990 and 2000 NYSDOT emissions factor tables for Nassau County. "Before" emissions factors for PM10 and PM2.5 are default 2000 values from the CMAQtraq program from NYSDOT.
Δ Total	NA	

Emissions factors used in calculations:

	BEFORE	AFTER
CO (g/mi)	15.79	0.13
VOC (g/mi)	3.14	0.05
NOx (g/mi)	25.54	16.62
PM2.5 (g/mi)	2.88	0.0061
PM10 (g/mi)	3.1570	---

COSTS

Project life:__4__ yrs   Interest rate:__7___%

	CMAQ	NON-CMAQ	TOTAL	METHODOLOGY/ASSUMPTIONS:
Capital	$1.0 M	$250,000	$1.25 M	The project life of the new CNG buses is 8 years. However, since comparison is against old buses which are near the end of their useful life, the cost-effectiveness analysis in this study accounts for fewer years. No cost effectiveness calculations were provided by the project sponsor.
Adm/oper	$0	$0	$0	
Total	$1.0 M	$250,000	$1.25 M	
Total annualized public cost:	$375,700			
Annual revenues:	None			
Net public cost:	$1.25 M			
Annual private cost	NA			
Total net cost	$1.25 M			

NOTE: Typically, replacement projects should only account for remaining useful life of the old buses, or the different in cost and emissions associated with a CNG bus vs. a conventional diesel bus.

Category: **DUST MITIGATION**	Subcategory: **Dust Mitigation**
CMAQ Project ID: CA20040439	Project Year: 2004
Location: Ridgecrest, California	MPO: Kern Council of Governments

Description: **Graaf Ave. Paving Project** - The "Graaf Avenue Paving Project" will provide funding for the City of Ridgecrest to pave 2 lanes of moving traffic, pave 2 lanes of parking, and install curb, gutter, and sidewalk on both sides of the street. This project includes the last unpaved section of Graaf Avenue within the city limits. It serves as a direct route to Immanual Christian School and other commercial activity centers. The total length of the project is 0.25 miles, along Graaf Avenue between Norma St. and Wayne St.

TRAVEL IMPACTS

		METHODOLOGY/ASSUMPTIONS:
△Vehicle trips:	NA	Using visual observation, there are about 300 average daily trips (ADT).
△VMT:	NA	
△Speed:	NA	The total project length is 1,320 feet.
△Delay:	NA	
△SOV	NA	27,375 annual VMT = 300 ADT * 365 days per year * (1,320 feet / 5,280 feet per mile)
△CP/VP	NA	
△Transit	NA	
△Walk	NA	
△Bike	NA	

EMISSIONS

		METHODOLOGY/ASSUMPTIONS:
△ VOC	NA	Using visual observation, the average traffic speed on the road is 30 mph. The silt content of the road material was estimated as 28.5%. The road carries residential traffic, comprised mostly of cars, SUVs, and trucks; therefore, the mean vehicle weight was estimated 2.5 tons. Data from the Western Regional Climate Center indicates the number of days with at least 0.01 inches of precipitation was 19 in 1998. The particle size multiplier was 0.36, from PM10 guidelines.
△ NOx	NA	
△ CO	NA	
△ PM10	- 143 kg/day	
△ PM2.5	NA	
△ Total	NA	

The following equation was used to calculate the quantity of size specific particulate emissions from the unpaved road.

$$E = 0.36(5.9)\left(\frac{28.5}{12}\right)\left(\frac{30}{30}\right)\left(\frac{2.5}{3}\right)^{0.7}\left(\frac{4}{4}\right)^{0.5}\left(\frac{365-19}{365}\right) \quad (1lb/VMT)$$

$$E = 4.22 \quad lb/VMT$$

COSTS

| | | | Project life:__20__ yrs | Interest rate: ___7__% |

	CMAQ	NON-CMAQ	TOTAL	METHODOLOGY/ASSUMPTIONS:
Capital	$174,360	$22,637	$197,360	Engineering costs were comprised of $16,909 CMAQ and $2,191 in local funding. Construction costs were split between $157,814 in CMAQ funding and $20,446 in local match.
Adm/oper	$0	$0	$0	
Total	$174,360	$22,637	$197,360	
Total annualized public cost:	$20,817			The cost effectiveness calculation provided by the project sponsor uses a 0.071 capital recovery factor.
Annual revenues:	None			(0.071 * $197,360) / 143 kg/day = $97.98 per kg of PM10
Net public cost:	$197,360			
Annual private cost	NA			
Total net cost	$197,360			

Category: **DUST MITIGATION**	Subcategory: **Dust Mitigation**
CMAQ Project ID: ID20040003	Project Year: 2004
Location: Sandpoint, Idaho	MPO: Bannock Planning Organization

Description: **Lincoln Ave. Sandpoint** - This project will fund the paving of Lincoln Avenue from Pine Street to Main Street in order to reduce the generation of PM10. Emissions inventory studies have shown that fugitive road dust emissions are a major source of PM10 emissions in most western US communities, contributing to 53% of the annual emissions and 37% of the winter-time daily emissions. The air quality improvement plan for the Sandpoint Nonattainment Area includes reducing fugitive road dust by paving and resurfacing.

TRAVEL IMPACTS

		METHODOLOGY/ASSUMPTIONS:
Δ Vehicle trips:	NA	Project length = 0.48 miles
Δ VMT:	NA	Average Daily Traffic (ADT) = 1,030 vehicles/day
Δ Speed:	NA	
Δ Delay:	NA	
Δ SOV:	NA	The project sponsor supplied ADT using vehicle counts. VMT was based on data
Δ CP/VP:	NA	obtained from Bannock Planning Organization 1998 Household Survey,
Δ Transit:	NA	COMPASS 1997-98 Valley Origin and Destination Study, and Northern Idaho
Δ Walk:	NA	Corridor Plans.
Δ Bike:	NA	

EMISSIONS

		METHODOLOGY/ASSUMPTIONS:
Δ VOC	NA	Particulate matter for unpaved roads = 0.360 kg/VMT
Δ NO$_X$	NA	Particulate matter for paved roads = 0.005 kg/VMT
Δ CO	NA	
Δ PM$_{10}$	- 175.512 kg/day	
Δ PM$_{2.5}$	NA	Emissions reduction = (0.360 kg/VMT – 0.005 kg/VMT) x (1,030 vehicles) x (0.48
Δ Total	-175.512 kg/day (0.193 tpd)	miles) = 175.512 kg/day

COSTS

		Project life: 20 yrs		Interest rate: 7 %	

	CMAQ	NON-CMAQ	TOTAL	METHODOLOGY/ASSUMPTIONS:
Capital	$319,600	$0	$319,600	Annual emissions reduction = 177 kg/day x 260 days = 46,020 kg/year emissions reduction
Adm/oper	$0	$0	$0	
Total	$319,600	$0	$319,600	
Total annualized public cost:	$33,710			The statewide average number of days with less than 0.01 inches precipitation is 260 days. 20-year project life was determined by the average life of maintained paved roadway in Idaho.
Annual revenues:	None			
Net public cost:	$319,600			
Annual private cost	NA			
Total net cost	$319,600			

NOTE: Emissions reductions for PM10 reported in the CMAQ database (255.11 kg/day) differ from those calculated and provided by the State (175 kg/day). A cost benefit calculation was provided by the State which assumed emissions benefits 260 days/year for 20 years.

Category: **DUST MITIGATION**	Subcategory: **Dust Mitigation**

CMAQ Project ID: ID20050017	Project Year: 2005
Location: Bannock County, Idaho	MPO: Bannock Planning Organization

Description: **Purchase of a Liquid De-Icer Truck** - Project to purchase a liquid de-icing truck to reduce application of sand and salt in the winter months thereby reducing PM10 emissions. The truck will use a combination of Magnesium Chloride on gravel roads and anti-icing chemicals on paved roads.

TRAVEL IMPACTS

		METHODOLOGY/ASSUMPTIONS:
ΔVehicle trips:	NA	Length of gravel road = 33.45 miles
ΔVMT:	141,845.5 /day	VMT on gravel = 2,247.3 miles/day
ΔSpeed:	NA	
ΔDelay:	NA	Length of paved road = 191.06 miles
ΔSOV	NA	VMT = 139,598.2 miles/day
ΔCP/VP	NA	
ΔTransit	NA	VMT was based on data obtained from Bannock Planning Organization 1998
ΔWalk	NA	Household Survey, COMPASS 1997-98 Valley Origin and Destination Study, and
ΔBike	NA	Northern Idaho Corridor Plans.

EMISSIONS

		METHODOLOGY/ASSUMPTIONS:
Δ VOC	NA	Emissions factors were obtained from the EPA's Compilation of Air Pollutant
Δ NO$_X$	NA	Emissions Factors (AP-42, September, 1998). Environmental staff at the Idaho
Δ CO	NA	Transportation Department determined the control efficiency was 0.70.
Δ PM$_{10}$	- 6,292 kg/day	
Δ PM$_{2.5}$	NA	*Reduction in PM10 by application of Magnesium Chloride* = 0.7073 PM emissions
Δ Total	- 6,292 kg/day (6.93 tpd)	factor * 0.70 * 2,247 VMT = 1,113 kg/day
		Reduction in PM10 by application of Anti-icing chemical = 0.053 PM emissions factor * 0.70 * 139,598 VMT = 5,179 kg/day
		Total PM10 reduction = 1,113 kg/day + 5,179 kg/day = 6,292 kg/day

COSTS

				Project life:__8_ yrs	Interest rate: __7 __%

	CMAQ	NON-CMAQ	TOTAL	METHODOLOGY/ASSUMPTIONS:
Capital	$152,889	$12,111	$165,000	The project life was determined by the average life of similar
Adm/oper	$0	$0	$0	equipment. Days of Activity per year = 200 days per year for
Total	$152,889	$12,111	$165,000	Magnesium Chloride and 90 days per year for the anti-icing
Total annualized public cost:	$29,865			chemicals
Annual revenues:	None			
Net public cost:	$165,000			
Annual private cost	NA			
Total net cost	$165,000			

NOTE: Costs include purchase of the trucks, but does not account for on-going operating costs. Assumption of 200 days of application per year sounds high, but is presumably accurate for the local area.

Category: **FREIGHT/ INTERMODAL**			Subcategory: **Freight/Intermodal**
CMAQ Project ID: ME20000004		Project Year: 2000	
Location: Cumberland and York Counties, Maine		MPO: Lewiston-Auburn Comprehensive Transportation Study	

Description: South Portland Truck to Rail Intermodal Facility - The South Portland Truck to Rail Intermodal Facility will provide funding to construct rail siding as part of an intermodal transfer. Inbound kaolin clay is currently transloaded from ships onto trucks for transport to paper mills. After completion of this project, the raw materials will be transported via rail, reducing the number of heavy duty vehicle trips required. The emissions analysis was conducted in 1999 and assumed the project would reach full capacity by 2006.

TRAVEL IMPACTS

Δ Vehicle trips:	- 2,250 /year	METHODOLOGY/ASSUMPTIONS:
Δ VMT:	- 225,000 /year	Assume the rate of usage is a constant 2,250 trucks per year by 2006 and one third of the trucks go North, one third South and one third West. Assume the average mileage per truck is 100 miles round trip and all truck emissions from heavy duty diesel vehicles (HDDV) traveling at 40 mph. The facility will operate 365 days per year.
Δ Speed:	NA	
Δ Delay:	NA	
Δ SOV	NA	
Δ CP/VP	NA	1999 750 trucks removed = 75,000 miles reduced
Δ Transit	NA	2006: 2,250 trucks removed = 225,000 miles reduced
Δ Walk	NA	2015: 2,250 trucks removed = 225,000 miles reduced
Δ Bike	NA	2018: 2,250 trucks removed = 225,000 miles reduced

EMISSIONS

Δ VOC	- 0.71 kg/day	METHODOLOGY/ASSUMPTIONS:
Δ NO$_x$	- 4.22 kg/day	Emissions reductions calculated using Mobile running emissions factors for HDDV at 40 mph. Emissions were calculated for 2006, 2015, and 2018 for comparison purposes.
Δ CO	NA	
Δ PM$_{10}$	NA	
Δ PM$_{2.5}$	NA	
Δ Total	- 4.93 kg/day (0.0054 tpd)	

COSTS

			Project life: __20__ yrs	Interest rate: ___7__%

	CMAQ	NON-CMAQ	TOTAL	METHODOLOGY/ASSUMPTIONS:
Capital	$283,941	$71,239	$355,180	
Adm/oper	$0	$0	$0	
Total	$283,941	$71,239	$355,180	
Total annualized public cost:	$41,096			
Annual revenues:	None			
Net public cost:	$355,180			
Annual private cost	NA			
Total net cost	$355,180			

NOTE: Analysis does not account for any increase in railroad emissions.

Category: **FREIGHT/ INTERMODAL**		Subcategory: **Freight/Intermodal**
CMAQ Project ID: ME20020005	Project Year: 2002	
Location: Cumberland and York Counties, Maine	MPO: Portland Area Comprehensive Transportation Study	

Description: **South Portland – Rail Line Rehab for Freight Shipping** - This project will fund the rehabilitation and/or replacement of tracks on the Sprague Industrial Spur. The rehab of the rail will allow freight to be shipped by rail instead of truck. Existing train traffic will not be impacted as additional freight will be added to existing trains. The project assuming that the existing truck traffic is traveling along the interstate south to reach Boston and beyond.

TRAVEL IMPACTS

		METHODOLOGY/ASSUMPTIONS:
Δ Vehicle trips:	- 1,000/ year	The number of trucks being removed from the highway is approximately 1,000 annually. Assumes that the trucks would normally travel down I-95 to Kittery traveling at an average speed 50 mph in Cumberland County and 60 mph in York County on the interstate. Assumes each round trip truck trip would be approximately 70 miles. Assume all truck emissions are from heavy duty diesel vehicles (HDDV) and the siding works 5 days a week 52 weeks a year.
Δ VMT:	- 70,000 /year	
Δ Speed:	NA	
Δ Delay:	NA	
Δ SOV	NA	
Δ CP/VP	NA	
Δ Transit	NA	2006: 1,000 trucks removed per year = 70,000 miles reduced
Δ Walk	NA	2015: 1,000 trucks removed per year = 70,000 miles reduced
Δ Bike	NA	2020: 1,000 trucks removed per year = 70,000 miles reduced

EMISSIONS

		METHODOLOGY/ASSUMPTIONS:
Δ VOC	- 0.18 kg/day	Emissions reductions calculated using Mobile running emissions factors for HDDV at 50 and 60 mph. Emissions were calculated for 2006, 2015, and 2018 for comparison purposes.
Δ NO$_x$	- 1.96 kg/day	
Δ CO	NA	
Δ PM$_{10}$	NA	
Δ PM$_{2.5}$	NA	
Δ Total	- 2.14 kg/day (0.0024 tpd)	

COSTS

			Project life:__20__ yrs	Interest rate: __7___%

	CMAQ	NON-CMAQ	TOTAL	METHODOLOGY/ASSUMPTIONS:
Capital	$128,501	$365,597	$494,098	
Adm/oper	$0	$0	$0	
Total	$128,501	$365,597	$494,098	
Total annualized public cost:	$54,720			
Annual revenues:	None			
Net public cost:	$494,098			
Annual private cost	NA			
Total net cost	$494,098			

NOTE: Analysis does not account for any increase in railroad emissions.

Category: **FREIGHT/INTERMODAL**	Subcategory: **Freight/Intermodal**
CMAQ Project ID: PA20020059, PA20030090	Project Year: 2002 - 2003
Location: Pittsburgh, Pennsylvania	MPO: Southwestern Pennsylvania Regional Planning Commission

Description: **Westmoreland Intermodal Freight Facility** - The Westmoreland Intermodal Center is a project to reduce the amount of freight cargo traveling through downtown Pittsburgh. A portion of the cargo currently entering and leaving the region on trucks will be diverted onto rail freight carriers at the Facility, reducing congestion on the region's freeways and major arterials. Currently, cargo either enters the region on trucks via major radial highways and is delivered to various points through the region or is picked-up at various points and leaves the region on trucks via the highways. By diverting a portion of the cargo to rail freight carriers serving the Intermodal Facility, the cargo will be transloaded to/from trucks at the facility for pick-up/delivery through the region.

TRAVEL IMPACTS

Δ Vehicle trips:	NA	METHODOLOGY/ASSUMPTIONS:
Δ VMT:	- 897/day	Assume 500,000 tons of cargo each year will be diverted to the Facility, the equivalent of 20,000 truck loads per year. Assume an average 1-way freight trip in the region without the facility is 94 miles, and with the facility will be 80 miles. Daily VMT reduction = 20,000 year truck trips / 312 operating days per year * 14 miles per trip = 897.4 VMT.

Calculate the change in truck VMT to deliver cargo with and without construction of the Facility. |
Δ Speed:	NA	
Δ Delay:	NA	
Δ SOV	NA	
Δ CP/VP	NA	
Δ Transit	NA	
Δ Walk	NA	
Δ Bike	NA	

EMISSIONS

Δ VOC	- 0.001 kg/day	METHODOLOGY/ASSUMPTIONS:
Δ NO$_x$	- 13.3 kg/day	Emissions reductions calculated using heavy duty diesel emissions factors from Mobile 5a_H emissions model for years 2001, 2004, and 2012 to compare emissions with and without construction of the facility. (Emissions change due to implementation of project = Yearly VMT x Emissions factor / 1000.)

Emissions reductions reported are for 2001 only. Travel speeds determined from the SPC travel demand model. Note that the average speed for a truck trip from the Intermodal Facility to the average pick-up/delivery point (38 mph) is lower than the average speed for a truck trip from a highway entering into the region to the average pick-up/delivery point (47 mph).
No Build Emissions @ 47.0 mph:
   VOC = 1,872,000 x 1.06 / 1000 = 1,982.32 kg/year
   NOx = 1,872,000 x 9.44 / 1000 = 17,671.68 kg/year
   CO = 1,872,000 x 5.24 / 1000 = 9,809.28 kg/year
Build Emissions @ 38.0 mph:
   VOC = 1,600,000 x 1.24 / 1000 = 1,984.00 kg/year
   NOx = 1,600,000 x 8.45 / 1000 = 13,520.00 kg/year
   CO = 1,600,000 x 5.76 / 1000 = 9,216.00 kg/year
Emissions Change = Build – No Build Emissions.
   VOC emissions change = 1,984.00 - 1,982.32 = -0.32 kg/year
   NOx emissions change = 13,520.00 – 17,671.68 = -4,151.68 kg/year
   CO emissions change = 9,216.00 – 9,809.28 = -593.28 kg/year |
Δ CO	- 1.90 kg/day	
Δ PM$_{10}$	NA	
Δ PM$_{2.5}$	NA	
Δ Total	-13.3 kg/day	

COSTS

				Project life:__20__ yrs	Interest rate: ___7___ %
	CMAQ	NON-CMAQ	TOTAL	METHODOLOGY/ASSUMPTIONS:	
Capital	$7.6 M	$1.9 M	$9.5 M	Assume trucks operate at the facility 6 days/week and the total service life of the project is 10 years. The cost-	
Adm/oper	$0	$0	$0		

Total	$7.6 M	$1.9 M	$9.5 M	effectiveness analysis in this study assumes a 20-year service life.
Total annualized public cost:	$1,028,000			
Annual revenues:	None			Cost-effectiveness calculations were not provided by sponsor.
Net public cost:	$9.5 M			
Annual private cost	NA			
Total net cost	$9.5 M			

NOTE: State provided information indicates total CMAQ funding of $7.6 M and a $1.9 M local match. The CMAQ database lists the project in FY 2002 ($8,750,000) and 2003 ($1,357,000) for a CMAQ funding total of $10,107,000. Analysis does not account for any increase in rail emissions. In the CMAQ database, kg/year figures were incorrectly reported as kg/day.

Category: **FREIGHT/INTERMODAL**	Subcategory: **Freight/Intermodal**

CMAQ Project ID: NY20040036	Project Year: 2004
Location: Staten Island, New York	MPO: Poughkeepsie-Dutchess County Transportation Council

Description: **Arlington Intermodal Yard** - This project will fund intermodal capacity improvements to the Staten Island Railroad's(SIRR) Arlington Yard. The project will improve the efficiency of rail shipments to and from Staten Island and the continental United States by increasing the reliability of the rail service and reduce the movement of freight shipments by truck through the NY metropolitan area. The project will extend rail track an additional 3,500 feet, construct a runaround track adjacent to the tail track, and construct a new right-hand crossover between the west end of the tail track and the runaround track. These seemingly minor improvements will greatly increase rail efficiencies at the yard by allowing one locomotive to be housed in the yard to classify and haul traffic for pickup or delivery.

TRAVEL IMPACTS

		METHODOLOGY/ASSUMPTIONS:
Δ Vehicle trips:	- 26,054 /day	The emissions calculation breaks the project into two segments: Visy Paper Mill – Bayoone Bridge and Transfer Station – Goethal Bridge.
Δ VMT:	- 140,538 /day	*Visy/Bayoone Segment*
Δ Speed:	NA	Average speed of 30 mph assumed for the transportation of freight by 10,268 heavy duty trucks per day. The average miles of vehicle travel per day is 6, for 302 days/year.
Δ Delay:	NA	Daily VMT Reduction = 10,268 trucks x 6 miles = 61,608.
Δ SOV	NA	Yearly VMT Reduction = 61,608 x 302 days = 18,605,616
Δ CP/VP	NA	*Trans/Goethal*
Δ Transit	NA	Average speed of 30 mph assumed for the transportation of freight by 15,786 heavy duty trucks per day. The average miles of vehicle travel per day is 5, for 302 days/year.
Δ Walk	NA	Daily VMT Reduction = 15,786 trucks x 5 miles = 78,930.
Δ Bike	NA	Yearly VMT Reduction = 78,930 x 302 days = 23,836,860.00
		Daily VMT Reduction = 23,836,860 / 302 days/year = 78,930 VMT reduction.

EMISSIONS

		METHODOLOGY/ASSUMPTIONS:
Δ VOC	- 209.01 kg/day	The NYSDOT software package CMAQtraq was used to estimate emissions for both segments, using the "Goods Vehicle Reduction" module. Effects were calculated for 302 days/year with the following emissions factors (g/mile):
Δ NO$_X$	- 1,008.80 kg/day	
Δ CO	- 1,712.21 kg/day	
Δ PM$_{10}$	- 37.00 kg/day	CO emissions factor = 12.18 PM2.5 emissions factor = 0.2184
Δ PM$_{2.5}$	- 30.07 kg/day	VOC emissions factor = 1.49 PM10 emissions factor = 0.2633
Δ Total	NA	NOx emissions factor = 7.18
		Total Emissions reduced = (Emissions factor * Visy/Bayoone miles reduced) + (Emissions factor * Trans/Goethal miles reduced)

COSTS

				Project life: 20_ yrs	Interest rate: ___7__ %

	CMAQ	NON-CMAQ	TOTAL	METHODOLOGY/ASSUMPTIONS:
Capital	$1.7 M	$7.3 M	$9.0 M	The NY Economic Development Commission will provide the local share of funding, as well as provide for the ancillary development costs. A cost effectiveness calculation was not provided by the project sponsor.
Adm/oper	$0	$0	$0	
Total	$1.7 M	$7.3 M	$9.0 M	
Total annualized public cost:	$949,300			
Annual revenues:	None			
Net public cost:	$9.0 M			

Annual private cost	NA	
Total net cost	$9.0 M	

NOTE: The $1.7 Million CMAQ funding total provided by the project sponsor does not match the $1.3 Million listed in the CMAQ database. The analysis seems to assume a very large number of trucks reduced, and does not account for rail emissions. It is not entirely clear if the truck assumptions were meant to be annual estimates rather than daily.

Category: **FREIGHT/INTERMODAL**	Subcategory: **Freight/Intermodal**
CMAQ Project ID: PA20040076	Project Year: 2004
Location: Indiana County, Pennsylvania	MPO: Southwestern Pennsylvania Commission MPO

Description: **Norfolk Southern Rail Extension and Rehabilitation** - This project will fund the construction of 5.25 miles and rehabilitation of 7 miles of rail track between Saltsburg and Shelocta, in Indiana County in order to create a more direct route for delivery of coal to the Keystone Power Station at Shelocta. Currently, coal is delivered by a combination of truck and rail freight movement. Increasing the amount of coal delivered via rail, will reduce the amount delivered by trucks. The new rail route is 106 miles less than the current route and has an easier grade, higher speed, and higher capacity than the existing rail line. Locomotive power required to haul 130 loaded coal cars per train over the new route will be less than that required to operate over the existing route with only 100 loaded coal cars per train; therefore, it is estimated that the new route will enable a decrease in the locomotive power required while, at the same time, increase by 30% the tonnage hauled per train.

TRAVEL IMPACTS

Δ Vehicle trips:	- 174 /day	METHODOLOGY/ASSUMPTIONS:
Δ VMT:	- 8,970 /day	Average Mine-based 2-way trip length = 41.26 miles
Δ Speed:	NA	Average Home-based 2-way trip Length is assumed to be 25% of the Mine-based length = 10.315 miles
Δ Delay:	NA	
Δ SOV	NA	Number of yearly 2-way truck trips before project = 107,396
Δ CP/VP	NA	Number of yearly 2-way truck trips after project = 63,918
Δ Transit	NA	Total Daily Truck VMT = Daily Mine-based VMT + Daily Home-based VMT
Δ Walk	NA	Daily Truck VMT Savings = VMT before project – VMT after project
Δ Bike	NA	VMT before = 5,538,967/year = 22,156/day; VMT after = 3,296,575/yr = 13,186/day
		Δ VMT = 8,970/day
		Assumes that ½ of the total coal delivered by truck to the power station originates in the project area, and consequently, ½ of the total truck VMT. Delivery occurs 250 days of the year. Estimated tons of coal that will be diverted from truck to rail = 1,000,000 tons/year.

EMISSIONS

Δ VOC	- 11.48 kg/day	METHODOLOGY/ASSUMPTIONS:
Δ NO$_x$	- 53.46 kg/day	Emissions reductions were calculated for heavy duty diesel trucks at an average speed of 36 mph using MOBILE6 emissions factors for 2004 and 2012.
Δ CO	- 64.67 kg/day	
Δ PM$_{10}$	NA	Reported emissions reductions are for 2004 only.
Δ PM$_{2.5}$	NA	Emissions factors:
Δ Total	-64.94 kg/day (0.0715 tpd)	VOC = 1.28; NOx = 7.21; CO = 5.96

COSTS

				Project life:__20__ yrs	Interest rate: ___7__%

	CMAQ	NON-CMAQ	TOTAL	METHODOLOGY/ASSUMPTIONS:
Capital	$10 M	$2.5 M	$12.5 M	Cost-effectiveness was not provided by sponsor.
Adm/oper	$0	$0	$0	
Total	$10 M	$2.5 M	$12.5 M	Annualized cost = $12.5 m x 0.081 CRF = $1.0125 mil
Total annualized public cost:	$1,318,500			C/E = $1.0125 mil / (0.0715 x 250) = $56,643/ton
Annual revenues:	None			
Net public cost:	$12.5 M			
Annual private cost	NA			
Total net cost	$12.5 M			

NOTE: The travel and emissions impact calculations did not include train emissions effects due to the increased tonnage of coal being carried by the rail cars.

Category **FREIGHT/INTERMODAL**	Subcategory: **Freight/Intermodal**
CMAQ Project ID: CT20060022	Project Year: 2006
Location: New Haven, Connecticut	MPO: South Central Regional COG

Description: Freight Rail Construction along Waterfront Street - This project will advance the railroad track installation on Waterfront St. in New Haven and the associated utility relocations. The track work will be performed by force account and the right of way for this work will be transferred from another project (92-541). The railroad track installation will improve air quality and reduce congestion by diverting some share of the transportation of freight and cargo through New Haven from truck to rail.

TRAVEL IMPACTS

		METHODOLOGY/ASSUMPTIONS:
Δ Vehicle trips:	- 15.38/day	Travel impact information was provided by the operator of the rail service in the area. Assumes 1,000 railcar shipments annually are diverted from truck to rail.
Δ VMT:	- 1,408 /day	Vehicle trips removed per year = 1,000 railcar shipments annually x 4 truck-equivalent per railcar / 260 days per year = 15.38 vehicle trips removed per day.
Δ Speed:	NA	
Δ Delay:	NA	
Δ SOV	NA	
Δ CP/VP	NA	50% of truck trips are one-way (61 miles average) and 50% of truck trips are round-trip
Δ Transit	NA	(122 miles average) from the Port of New Haven.
Δ Walk	NA	Daily VMT reduction = (50% x 4000 annual truck trips x 61 miles / 260 days/year) + (50% x 4000 truck trips x 122 miles / 260 days/year) = 1,408 VMT reduced daily.
Δ Bike	NA	
		In the $PM_{2.5}$ non-attainment areas, 50% of truck trips are one-way (27 miles average) and 50% of truck trips are round trip (43 miles average).
		Daily VMT reduction = (50% x 4000 annual truck trips x 27 miles / 260 days/year) + (50% x 4000 truck trips x 54 miles / 260 days/year) = 623 VMT reduced daily in $PM_{2.5}$ non-attainment areas.

EMISSIONS

		METHODOLOGY/ASSUMPTIONS:
Δ VOC	- 0.460 kg/day	Assumes 90% of a typical truck trip is on expressway and 10% is on arterial.
Δ NO$_x$	- 18.437 kg/day	*Emissions factors, expressway:*
Δ CO	NA	VOC = 0.314 g/mile; NOx = 13.630 g/mile; $PM_{2.5}$ = 0.261 g/mile
Δ PM$_{10}$	NA	*Emissions factors, arterial:*
Δ PM$_{2.5}$	- 0.162 kg/day	

Δ Total	NA	VOC = 0.453 g/mile; NOx = 8.300 g/mile; $PM_{2.5}$ = 0.260 g/mile
		VOC emissions reduction = (1,408 VMT reduction x 0.314 g/mile x 90%) + (1,408 VMT reduction x 0.453 g/mile x 10%) = 461 g/day VOC reduction NOx emissions reduction = (1,408 VMT reduction x 13.630 x 90%) + (1,408 VMT reduction x 8.3 g/mile x 10%) = 18,441 g/day NOx reduction $PM_{2.5}$ emissions reduction = (623 VMT reduction x 0.261 g/mile x 90%) + (623 VMT reduction x 0.260 g/mile x 10%) = 162 g/day $PM_{2.5}$ reduction Additional locomotive emissions due to increased train trip length to service the additional trackage (Additional emissions = gallons consumed x emissions factor): Estimate 15 gallons of additional fuel consumed per day. Use EPA emissions factors, based on the age of the locomotive fleet. VOC = 15 gallons x 21.0 g/gal = 0.32 g/day of additional VOC emissions. NOx = 15 gallons x 262.0 g/gal = 3.93 g/day of additional NOx emissions. $PM_{2.5}$ = 15 gallons x 9.2 g/gal = 0.14 g/day of additional $PM_{2.5}$ emissions. Overall emissions reductions: VOC emissions reduction = 461 – 0.32 = 460.68 g/day NOx emissions reduction = 18,441 – 3.93 = 18,437.07 g/day $PM_{2.5}$ emissions reduction = 162 – 0.14 = 161.86 g/day

COSTS				
Annualized public costs			Project life:__20__ yrs	Interest rate: ___7__%
	CMAQ	NON-CMAQ	TOTAL	METHODOLOGY/ASSUMPTIONS: Assumes project has movement 260 days/year.
Capital	$1,409,600	$352,400	$1,762,000	
Adm/oper	$0	$0	$0	Cost-effectiveness was not provided by the project sponsor.
Total	$1,409,600	$352,400	$1,762,000	
Total annualized public cost:		$174,100		
Annual revenues:		None		
Net public cost:		$1,762,000		
Annual private cost		NA		
Total net cost		$1,762,000		

NOTE: Sponsor's calculation incorrectly showed value in kg rather than grams, overstating emissions effects. The analysis could also consider what portion of the truck's VMT is occurring in the impact area.

Category **DIESEL EMISSIONS REDUCTION**	Subcategory: **Diesel Engine Retrofits**
CMAQ Project ID: MD20010025	Project Year: 2001
Location: Baltimore, Maryland	MPO: Baltimore Metropolitan Council

Description: 12 Bus Engine Upgrade - This project will fund 142 engine overhauls on the MTA bus fleet, including the installation of catalytic converters using an EPA certified engine rebuild kit. Each bus engine's rebuild is scheduled as needed or after 300,000 miles of service.

TRAVEL IMPACTS

		METHODOLOGY/ASSUMPTIONS:
Δ Vehicle trips:	NA	
Δ VMT:	NA	
Δ Speed:	NA	MTA has 794 buses available for service and the total vehicle miles in FFY 2001 were 21,774,843.
Δ Delay:	NA	
Δ SOV	NA	
Δ CP/VP	NA	(21,774,843 miles / 794 buses * 142 Buses = 3,894,241 overhaul bus miles)
Δ Transit	NA	
Δ Walk	NA	
Δ Bike	NA	

EMISSIONS

		METHODOLOGY/ASSUMPTIONS:
Δ VOC	NA	Southwestern Research and MTA data indicates that each gram per brake horsepower-hour (g/bhp-hr) is equal to 1.5 tons of PM per 100,000 bus-miles.
Δ NO$_x$	NA	
Δ CO	NA	
Δ PM$_{10}$	NA	
Δ PM$_{2.5}$	- 34.77 kg/day	(3,894,241 overhaul bus miles / 100,000 miles * 1.5 tons = 58.41 tons of PM per g/bhp-hr)
Δ Total	NA	

Data provided by the Engelhard Automotive Emissions Systems indicates that depending on the age of a bus and engine type, the PM can be reduced between 0.46 and 0.05 g/bhp-hr. The average will vary around 0.255 g/bhp-hr of PM.

0.255 g/bhp-hr x 58.41 (conversion of g/bhp-hr to tons)= 14.9 tons of PM per year

14.9 tons of PM per year * 2.4837895 conversion factor = kilograms / day

COSTS

| | | | Project life:_7_ yrs | Interest rate: ___7__% |

	CMAQ	NON-CMAQ	TOTAL	METHODOLOGY/ASSUMPTIONS:
Capital	$5.458 M	$17.578 M	$23.036 M	The total cost budgeted for the MTA fleet is $23.036 million; funds requested from the CMAQ program total $5.458 million, or 23.69 % of the total cost. Cost-effectiveness was not provided by the project sponsor.
Adm/oper	$0	$0	$0	
Total	$5.458 M	$17.578 M	$23.036 M	
Total annualized public cost:	$5,095,000			
Annual revenues:	None			
Net public cost:	$23.036 M			
Annual private cost	NA			
Total net cost	$23.036 M			

NOTE: The CMAQ funding portion listed in the CMAQ database for this project is $4,366,000.

Category: **DIESEL EMISSIONS REDUCTION**	Subcategory: **Diesel Engine Retrofits**
CMAQ Project ID: NY20040032	Project Year: 2004
Location: New York	MPO: New York Metropolitan Transportation Council

Description: **WCDOT Diesel Engine Retrofit of 177 Transit Buses** - This project will fund the retrofit of 177 Bee-Line buses with USEPA verified Englehard DPX filters, a class of equipment known as passive regenerative catalyzed diesel PM filters. The fleet consists of 99 standard 40' Orion buses and 78 Neoplan articulated buses.

TRAVEL IMPACTS

Δ Vehicle trips:	NA	**METHODOLOGY/ASSUMPTIONS:**
Δ VMT:	NA	Orion buses travel 10,944 weekday miles and 9,784 weekend miles. Neoplan buses travel 7,056 weekday miles and 9,500 weekend miles. This schedule reflects peak vehicle use and includes both revenue miles and dead-head miles provided by the Westchester DOT and the service contractor.
Δ Speed:	NA	
Δ Delay:	NA	
Δ SOV	NA	
Δ CP/VP	NA	
Δ Transit	NA	Assume 254 weekdays per year and 104 weekend days per year.
Δ Walk	NA	
Δ Bike	NA	

EMISSIONS

Δ VOC	- 2.96 kg/day	**METHODOLOGY/ASSUMPTIONS:**
Δ NOx	+ 14.64 kg/day*	Emissions reductions were calculated from emissions with the retrofits provided by the Transit Resource Center and existing emissions from tailpipe testing. The average speed from testing Bee-Line bus duty cycle is 20 mph.
Δ CO	- 45.94 kg/day	
Δ PM$_{10}$	NA	
Δ PM$_{2.5}$	NA	
Δ Total	NA	

Testing Condition	Emissions Factors (g/mi)			
Orion:	**CO**	**VOC**	**NOx**	**PM**
Existing:	7.94	0.31	39.4	0.62
Retrofit:	0.36	0.0	38.5	0.062
Neoplan:	**CO**	**VOC**	**NOx**	**PM**
Existing:	1.59	0.34	31.4	0.29
Retrofit:	0.11	0.0	36.4	0.034

Total emissions reduced = (Number of Orion buses x Orion Bus VMT x (Existing emissions rate – retrofit emissions rate)) + (Number of Neoplan buses x Neoplan Bus VMT x (Existing emissions rate – retrofit emissions rate))

COSTS

Project life:_7_ yrs   Interest rate: ___7__%

	CMAQ	NON-CMAQ	TOTAL	**METHODOLOGY/ASSUMPTIONS:**
Capital	$1.2 M	$300,000	$1.5 M	
Adm/oper	$0	$0	$0	
Total	$1.2 M	$300,000	$1.5 M	
Total annualized public cost:	$311,000			
Annual revenues:	None			
Net public cost:	$1.5 M			
Annual private cost	NA			
Total net cost	$1.5 M			

NOTE: The project sponsor noted that the increase in NOx emissions found is highly unusual, and it is widely accepted that this technology will have no impact on NOx. The other reductions also do not match with EPA's estimates: 60% reduction in HC, CO, and PM; no impact on NOx. The project sponsor did not calculate PM emissions reductions, but the cost-effectiveness analysis determined these values should be (2.3 kg/day reduced) The CMAQ funding portion listed in the CMAQ database for this project is $4,366,000.

Category: **DIESEL EMISSIONS REDUCTION**	Subcategory: **Diesel Engine Retrofits**
CMAQ Project ID: PA20040011	Project Year: 2004
Location: Philadelphia, Pennsylvania	MPO: Delaware Valley Regional Planning Commission

Description: **Install 235 Emissions Reduction Device on Local Buses** - This project includes the purchase and installation of 235 emissions reduction devices in SEPTA's bus fleet. The proposed retrofit device is an Englehard DPX soot filter, an exhaust emissions filter. 155 buses are Neoplan Articulated buses with a 1999 Detroit Diesel Series 50 engine and mileages ranging from 70,942 to 110,361 miles. 80 of the buses are Eldorado buses with a 2000 Cummins ISB engine and mileages ranging from 38,733 to 149,199 miles.

TRAVEL IMPACTS

Δ Vehicle trips:	NA	**METHODOLOGY/ASSUMPTIONS:**
Δ VMT:	NA	The average annual vehicle revenue miles for each of the 235 buses is 27,207 miles, with a deadhead factor of 1.15, and an average bus speed of 13.5 mph. Assumes each bus operates an average of 250 days per year.
Δ Speed:	NA	
Δ Delay:	NA	
Δ SOV	NA	
Δ CP/VP	NA	
Δ Transit	NA	
Δ Walk	NA	
Δ Bike	NA	

EMISSIONS

Δ VOC	-7.33 kg/day	**METHODOLOGY/ASSUMPTIONS:**
Δ NO$_x$	NA	The EPA has certified this device to reduce HC and CO emissions by 60%, with no impact on NOx. Emissions reductions were calculated by multiplying the total emissions for the 1999 and 2000 buses using MOBILE6 emissions factors and then multiplying the VOC and CO emissions by 0.60.
Δ CO	-111.22 kg/day	
Δ PM$_{10}$	NA	
Δ PM$_{2.5}$	NA	
Δ Total	-7.33 kg/day (0.008 tpd)	

COSTS

Project life: __7-8__ yrs   Interest rate: __7___%

	CMAQ	NON-CMAQ	TOTAL	METHODOLOGY/ASSUMPTIONS:
Capital	$1,793,520	$449,000	$2,242,520	
Adm/oper	$0	$0	$0	Selected through the 2002 DVRPC Competitive CMAQ Program by the MPO, for funding in 2004. Cost-effectiveness was not provided by sponsor.
Total	$1,793,520	$449,000	$2,242,520	
Total annualized public cost:	NA			
Annual revenues:	None			
Net public cost:	$2,242,520			
Annual private cost:	NA			
Total net cost	$2,242,520			

NOTE: Emissions reductions provided by the State sponsor do not match those reported in the CMAQ database. Reported values are – 0.77 VOC, - 6.29 CO, and - 0.57 NOx. EPA reports a 60% reduction in PM emissions, which are included in the cost-effectiveness calculations as 6.5 kg/day PM$_{10}$ reduction and 5.7 kg/day PM$_{2.5}$.

Category: **DIESEL EMISSIONS REDUCTION**			Subcategory: **Diesel Engine Retrofits**	
CMAQ Project ID: OR20050011			Project Year: 2005	
Location: Medford, Oregon			MPO: Rogue Valley COG	

Description: Install Filters on 9 Trash Collection Vehicles – This project will fund the installation of advanced exhaust after-treatment controls on 9 diesel-powered trash collection trucks owned by Rogue Disposal and Recycling refuse collection company. The EPA's Retrofit Calculator was used to identify those 9 vehicles in the garbage hauling fleet's trucks that would be most cost effectively fitted with these devices. The filters will effectively reduce emissions of diesel particulate by over 80 percent, while also reducing CO and VOC emissions by 67 percent and 95 percent, respectively. Additional toxin reductions will include benzene, formaldehyde, and polycyclic aromatic hydrocarbons.

TRAVEL IMPACTS

		METHODOLOGY/ASSUMPTIONS:
Δ Vehicle trips:	NA	
Δ VMT:	NA	
Δ Speed:	NA	Rogue Disposal's fleet of eighteen diesel-powered trash collection trucks consume
Δ Delay:	NA	over 400,000 gallons of fuel per year, emit a total of 2.13 tons of CO, 0.86 tons of
Δ SOV	NA	VOCs, and 0.2 tons of PM, and travel up to 855,000 total miles annually.
Δ CP/VP	NA	
Δ Transit	NA	
Δ Walk	NA	
Δ Bike	NA	

EMISSIONS

		METHODOLOGY/ASSUMPTIONS:
Δ VOC	- 1.43 kg/day	The Rogue Valley MPO provided Mobile6 inputs for EPA's Retrofit Calculator.
Δ NO$_X$	NA	Based on the manufacturer's report, the following manual calculations were used:
Δ CO	- 2.49 kg/day	
Δ PM$_{10}$	- 0.28 kg/day	VOC reductions: 0.86 tons VOC per year x 9/18 x 95% reduction x 907 conversion
Δ PM$_{2.5}$	NA	factor / 260 days = 1.43 kg reduced per day
Δ Total	-1.7 kg/day	
		CO reductions: 2.13 tons VOC per year x 9/18 x 67% reduction x 907 conversion factor / 260 days = 2.49 kg reduced per day
		PM reductions: 0.2 tons VOC per year x 9/18 x 80% reduction x 907 conversion factor / 260 days = 0.28 kg reduced per day

COSTS

Project life:__7__ yrs   Interest rate: ___7__%

	CMAQ	NON-CMAQ	Total	METHODOLOGY/ASSUMPTIONS:
				The project received an increase in funds to retrofit nine additional vehicles in July 2007. The grand total of this project has been amended to $124,615. CMAQ is funding 80% of this project, $99,692.
Capital	$49,692	$12,423	$62,115	
Adm/oper	$0	$0	$0	
Total	$49,692	$12,423	$62,115	
Total annualized public cost:	$12,457			
Annual revenues:	None			
Net public cost:	$62,115			
Annual private cost	NA			
Total net cost	62,115			

NOTE: Emissions reductions calculated above slightly differ from values reported in the CMAQ database (VOC: 1.39 kg/day, CO: 2.31 kg/day, PM$_{10}$: 0.28 kg/day). EPA's retrofit calculator has since been replaced by new modeling tools: the Diesel Emissions Quantifier and the National Mobile Inventory Model (NMIM).

Category: **DIESEL EMISSIONS REDUCTION**	Subcategory: **Diesel Engine Retrofits**
CMAQ Project ID: Not Yet Assigned	Project Year: 2007
Location: Detroit, Michigan	MPO: Southeast Michigan Council of Governments

Description: **Locomotive Diesel Engine Retrofits** - This project will fund the repowering of six switcher (4-axle) locomotives at the CSXT Rougemere Rail Yard with ultra-clean Generator Set (GENSET) diesel locomotive engine technology. The repowered locomotives will operate within the CSXT Detroit rail yard for five years. Each existing locomotive is stripped from the deck up, removing the large, single diesel engine that provides power. The engine is replaced with three smaller, ultra-clean diesel generators, which are fitted onto the platform, along with new control, and operating equipment. The Southeast Michigan Council of Governments (SEMCOG) anticipates repowering three locomotives in 2007 and another three in 2008; this analysis only accounts for the three 2007 repowerings.

TRAVEL IMPACTS

		METHODOLOGY/ASSUMPTIONS:
△Vehicle trips:	NA	None.
△VMT:	NA	
△Speed:	NA	
△Delay:	NA	
△SOV	NA	
△CP/VP	NA	
△Transit	NA	
△Walk	NA	
△Bike	NA	

EMISSIONS

		METHODOLOGY/ASSUMPTIONS:
△ VOC	- 9.96 kg/day	Baseline NOx emissions are calculated using the USEPA emissions standard for a Tier 0 switch engine. Baseline PM and VOC emissions are conservatively calculated using the lower uncontrolled USEPA emissions factors of 0.44 g/bhp-hr PM and 1.01 g/bhp-hr VOC in lieu of the Tier 0 standards of 0.72 g/bhp-hr PM and 2.1 g/bhp-hr VOC. Conventional switcher fuel use is 60,000 gal/yr. According to USEPA, the Brake-Specific Fuel Consumption (BSFC) is 20.8 bhp-hr/gal. GENSET locomotive emissions factors are based on Tier 3 nonroad engine standards (grams/bhp-hr) since the locomotive is powered by these engines. NMHC has the same meaning as VOC and HC. The GENSET fuel use is 25% less than a conventional switcher (45,000 gal/yr). The Brake-Specific Fuel Consumption (BSFC) for a GENSET engine is reported as 19.5 bhp-hr/gal.
△ NOx	- 132.1 kg/day	
△ CO	NA	
△ PM$_{10}$	NA	
△ PM$_{2.5}$	- 3.68 kg/day	
△ Total	NA	

NOX Emissions (kg/yr) = BSFC * Fuel Used * Emissions Factor * 3 Locomotives
   Baseline NOx = 20.8 * 60,000 * 14.0 * 3 = 52,416 kg/year
   GENSET NOx = 19.5 * 45,000 * 2.85 * 3 = 7,503 kg/year
   Daily reduction in NOx = (52,416 - 7,503) / 340 days per year = 132.1 kg/day

Calculation also assumes 86% reduction in ozone precursors and 76% reduction in PM.

COSTS

		Project life:__5__ yrs		Interest rate: ___7__%

	CMAQ	NON-CMAQ	TOTAL	METHODOLOGY/ASSUMPTIONS:
Capital	$3.36 M	$840,000	$4.2 M	A separate emissions and cost effectiveness analysis was conducted for the three locomotives scheduled for repowering in 2008. Assumes the three locomotives operate 340 days per year with a 5 year design life.
Adm/oper	$0	$0	$0	
Total	$3.36 M	$840,000	$4.2 M	
Total annualized public cost:	$1,042,000			Cost effectiveness = Total project cost / (Emissions reduction * 5 years * 340 days)
Annual revenues:	None			Cost per Kg over the design life for VOC = $248.05
Net public cost:	$4.2 M			Cost per Kg over the design life for NOx = $18.70
Annual private cost:	NA			Cost per Kg over the design life for PM2.5 = $671.36
Total net cost	$4.2 M			

NOTE: The VOC and PM$_{1.5}$ emissions reductions were calculated using similar methods as the NOx calculation shown.

Category: **DIESEL EMISSIONS REDUCTION**		Subcategory: **Diesel Engine Retrofits**
CMAQ Project ID: Not Yet Assigned		Project Year: 2007
Location: Orangetown, New York		MPO: New York Metropolitan Transportation Council

Description: **Diesel Engine Retrofits of 53 County Vehicles** - This project will fund the diagnostic review and installation of retrofit devices on all on-road diesel fueled vehicles over 8,500 pounds that are owned by the City of Orangetown. Each vehicle type will require a different treatment and diagnostic review to ensure that the retrofits meet the needs of the vehicle's documented usage and performance. All of the retrofit technologies will be verified by the US EPA or the California Air Resources Board (CARB).

TRAVEL IMPACTS

		METHODOLOGY/ASSUMPTIONS:
△ Vehicle trips:	NA	This project will retrofit approximately 53 on-road diesel powered vehicles and heavy duty equipment:
△ VMT:	NA	
△ Speed:	NA	HDDV8A: 25 Vehicles
△ Delay:	NA	HDDV8B: 6 Vehicles
△ SOV	NA	HDDV7: 14 Vehicles
△ CP/VP	NA	HDDV6: 1 Vehicle
△ Transit	NA	HDDV5: 2 Vehicles
△ Walk	NA	HDDV4: 1 Vehicle
△ Bike	NA	HDDV3: 4 Vehicles

EMISSIONS

		METHODOLOGY/ASSUMPTIONS:
△ VOC	- 0.44 kg/day	Emissions reductions calculated using Mobile6, EPA, and CARB emissions factors for a variety of heavy duty vehicles traveling at 23 mph for 4.43 miles per day for 305 days/year with and without the retrofits. The assumed speed, operating days, and travel distance was reported by Rockland County Highway Dept. as an average of their fleet vehicles equipped with GPS technology.
△ NO_X	NA	
△ CO	-1.71 kg/day	
△ PM_{10}	- 0.18 kg/day	
△ $PM_{2.5}$	- 0.14 kg/day	
△ Total	NA	

COSTS

				Project life:__7__ yrs	Interest rate: ___7__%

	CMAQ	NON-CMAQ	TOTAL	METHODOLOGY/ASSUMPTIONS:
Capital	$424,000	$106,000	$530,000	Sponsor assumes $10K/vehicle/retrofit for all 53 vehicles. A constant flow of benefits were assumed for 7 years following project completion. Cost effectiveness was calculated as kg of pollution reduced for each CMAQ dollar of funding.
Adm/oper	$0	$0	$0	
Total	$424,000	$106,000	$530,000	
Total annualized public cost:	$100,100			CO effectiveness: - 1.23 kg per CMAQ$(thousands)
Annual revenues:	None			VOC effectiveness: - 0.32 kg per CMAQ$(thousands)
Net public cost:	$530,000			PM2.5 effectiveness: - 0.10 kg per CMAQ$(thousands)
Annual private cost	NA			PM10 effectiveness: - 0.13 kg per CMAQ$
Total net cost	$530,000			

NOTE: Benefits were estimated for 7 years (2011-2017). The calculation only assumes the vehicles travel 4.43 miles per day per vehicle, which sounds conservative.

Category: **DIESEL EMISSIONS REDUCTION**		Subcategory: **Diesel Engine Retrofits**
CMAQ Project ID: Not Yet Assigned		Project Year: 2007
Location: Rockland County, New York		MPO: New York Metropolitan Transportation Council

Description: **3 Locomotive Repowers** - This project will fund the diagnostic review and installation of retrofit devices on 134 on-road diesel powered vehicles, public transit buses, and DPW vehicles. Rockland County has passed a law requiring all County-owned on-road diesel fueled vehicles that have a gross weight over 8,500 pounds to be retrofitted with the best available technology verified by EPA or CARB. This project is a direct result of that law, though only vehicles with replacement years of 2010 and beyond will be retrofitted with the project.

TRAVEL IMPACTS

		METHODOLOGY/ASSUMPTIONS:
ΔVehicle trips:	NA	
ΔVMT:	NA	
ΔSpeed:	NA	
ΔDelay:	NA	
ΔSOV	NA	
ΔCP/VP	NA	
ΔTransit	NA	
ΔWalk	NA	
ΔBike	NA	

EMISSIONS

		METHODOLOGY/ASSUMPTIONS:
Δ VOC	- 140.33 kg/day	Emissions reductions calculated using Mobile6, EPA, and CARB emissions factors for a variety of vehicle types traveling at posted speeds. The assumed speed, operating days, and travel distance was reported by Rockland County Highway Dept. as an average of their fleet vehicles equipped with GPS technology.
Δ NO$_X$	NA	
Δ CO	- 969.84 kg/day	
Δ PM$_{10}$	- 138.35 kg/day	
Δ PM$_{2.5}$	- 125.88 kg/day	
Δ Total	NA	

COSTS

Project life:__7-15__ yrs   Interest rate: ___7__ %

	CMAQ	NON-CMAQ	TOTAL	METHODOLOGY/ASSUMPTIONS:
Capital	$1.368 M	$342,000	$1.71 M	Sponsor assumes $10K/vehicle/retrofit for all vehicles. A constant flow of benefits were assumed for 7-15 years following project completion. Cost effectiveness was calculated as kg of pollution reduced for each CMAQ dollar of funding.
Adm/oper	$0	$0	$0	
Total	$1.368 M	$342,000	$1.71 M	
Total annualized public cost:	$323,000			CO effectiveness: - 242.21 kg per CMAQ$
Annual revenues:	None			VOC effectiveness: - 34.92 kg per CMAQ$
Net public cost:	$1.71 M			PM2.5 effectiveness: - 31.03 kg per CMAQ$
Annual private cost	NA			PM10 effectiveness: - 34.10 kg per CMAQ$
Total net cost	$1.71 M			

NOTE: The cost-effectiveness analysis used in this study accounts for a stream of benefits over 7 years.

Category: **DIESEL EMISSIONS REDUCTION**	Subcategory: **Truck Idle Reduction**
CMAQ Project ID: TN20030011	Project Year: 2003
Location: Knoxville, Tennessee	MPO: Knoxville Urbanized Area MPO

Description: **TSE: IdleAire 100 Units at Watt Rd.** - This project will provide funding for the installation of 100 IdleAire units at various truck stops at InterState 40 and Watt Rd. The IdleAire devices are capable of providing heating, air-conditioning, and other services to a truck cab, as an alternative to using the truck's engine to provide continuous power. Emissions reductions will be achieved due to the fact that the trucks no longer have to idle their engines in order to have access to heating or air-conditioning.

TRAVEL IMPACTS

		METHODOLOGY/ASSUMPTIONS:
Δ Vehicle trips:	NA	None.
Δ VMT:	NA	
Δ Speed:	NA	
Δ Delay:	NA	
Δ SOV	NA	
Δ CP/VP	NA	
Δ Transit	NA	
Δ Walk	NA	
Δ Bike	NA	

EMISSIONS

		METHODOLOGY/ASSUMPTIONS:
Δ VOC	- 4.47 kg/day	Emissions reductions were calculated using a truck idle emissions factor from MOBILE6. Emissions for one truck idling for 10 hours each day are estimated to be removed by each IdleAire unit, as reported by the IdleAire company.
Δ NO$_x$	- 60.4 kg/day	
Δ CO	NA	
Δ PM$_{10}$	NA	
Δ PM$_{2.5}$	NA	
Δ Total	- 64.87 kg/day	VOC emissions reduction = 44.7 g/unit * 100 units / 1000 = 4.47 kg/day NOx emissions reduction = 604 g/unit * 100 units / 1000 = 60.4 kg/day

COSTS

Project life:__10__ yrs   Interest rate: ___7__%

	CMAQ	NON-CMAQ	TOTAL	METHODOLOGY/ASSUMPTIONS:
Capital	$1.0 M	$0	$1.0 M	Cost-effectiveness was not provided by the project sponsor.
Adm/oper	$0	$0	$0	
Total	$1.0 M	$0	$1.0 M	
Total annualized public cost:	$163,300			
Annual revenues:	None			
Net public cost:	$1,000,000			
Annual private cost	NA			
Total net cost	$1,000,000			

Note: When the figures were recalculated using EPA's current guidance on long duration truck idling; this results in a 135.0 kg/day reduction of NOx and 3.68 kg/day reduction of PM.

Category: **DIESEL EMISSIONS REDUCTION**	Subcategory: **Truck Idle Reduction**
CMAQ Project ID: KY20060013	Project Year: 2006
Location: Oak Grove, Kentucky	MPO: Clarksville MPO

Description: **Advance Travel Center Electrification (IdleAire)** - The City of Oak Grove hired IdleAire to install 50 ATE units at Pilot Travel Center #49, a site conveniently located off of I-24 at Exit 89, in Christian County, Kentucky. These IdleAire units provide heating and air conditioning, as well as a wide range of communication and entertainment options, directly into the cab of a parked truck. This allows a truck driver to completely shut down the truck's engine, eliminating the air pollution associated with idling. This such ATE project will save 182,500 gallons of diesel fuel annually, remove over 1,700 tons of emissions annually, and remove existing barriers to the quality of driver rest, such as noise, vibrations, and fumes.

TRAVEL IMPACTS

		METHODOLOGY/ASSUMPTIONS:
Δ Vehicle trips:	NA	
Δ VMT:	NA	
Δ Speed:	NA	
Δ Delay:	NA	
Δ SOV	NA	
Δ CP/VP	NA	
Δ Transit	NA	
Δ Walk	NA	
Δ Bike	NA	

EMISSIONS

		METHODOLOGY/ASSUMPTIONS:
Δ VOC	-6.68 kg/day	The emission factors for CO and VOC come from EPA's Mobile6 Emissions Model to estimate the emissions from idling trucks. NOx and PM factors are calculated based on 2004 EPA Guidance.
Δ NO$_X$	-110.17 kg/day	
Δ CO	- 46.74 kg/day	
Δ PM$_{10}$	Na	
Δ PM$_{2.5}$	Na	Emissions Reduction = Emission Factor (g/unit) x 50 units / 1000 (g to kg conversion factor) = Emissions Reduction (kg/day).
Δ Total	Na	An IdleAire utilization rate of 10 hours per day / 365 days per year is assumed.

COSTS

				Project life:__15-20__ yrs	Interest rate: ___7__ %

	CMAQ	NON-CMAQ	TOTAL	METHODOLOGY/ASSUMPTIONS:
Capital	$500,000	$355,000	$835,000	The CMAQ funds will satisfy the capital needs of the initial installation of these units. The IdleAire Corp will pay for the on-going operational costs.
Adm/oper	$0	$0	$0	
Total	$500,000	$355,000	$835,000	
Total annualized public cost:	NA			
Annual revenues:	None			
Net public cost:	$835,000			
Annual private cost	Na			
Total net cost	$835,000			

Category: **DIESEL EMISSIONS REDUCTION**			Subcategory: **Truck Idle Reduction**		
CMAQ Project ID: TN20060026			Project Year: 2006		
Location: Jefferson County, Tennessee			MPO: Knoxville Urbanized Area MPO		
Description: **TSE: IdleAire Units in Jefferson Co.** - This project will provide funding for the installation of 59 IdleAire units at one travel center truck stop in Jefferson County. The IdleAire devices are capable of providing heating, air-conditioning, and other services to a truck cab, as an alternative to using the truck's engine to provide continuous power. Emissions reductions will be achieved due to the fact that the trucks no longer have to idle their engines in order to have access to heating or air-conditioning.					

TRAVEL IMPACTS					
Δ Vehicle trips:	NA	METHODOLOGY/ASSUMPTIONS:			
Δ VMT:	NA	None.			
Δ Speed:	NA				
Δ Delay:	NA				
Δ SOV	NA				
Δ CP/VP	NA				
Δ Transit	NA				
Δ Walk	NA				
Δ Bike	NA				

EMISSIONS					
Δ VOC	NA	METHODOLOGY/ASSUMPTIONS:			
Δ NO$_x$	-79.65 kg/day	Idle emissions factors for trucks are taken from the EPA guidance document – *"Guidance for Quantifying and Using Long Duration Truck Idling Emissions Reductions in State Implementation Plans and Transportation Conformity."*			
Δ CO	NA				
Δ PM$_{10}$	NA				
Δ PM$_{2.5}$	- 2.17 kg/day	Emissions for one truck idling for 10 hours each day are estimated to be removed by each IdleAire unit, as reported by the IdleAire company.			
Δ Total	- 81.8 kg/day				
		NOx emissions reduction = 135 g/hr/unit * 10 hours * 59 units = 79.65 kg/day			
		PM$_{2.5}$ emissions reduction = 3.68 g/hr/unit * 10 hours * 59 units = 2.17 kg/day			

COSTS					
			Project life: __10__ yrs		Interest rate: ___7__%
	CMAQ	NON-CMAQ	TOTAL	METHODOLOGY/ASSUMPTIONS:	
Capital	$788,240	$197,060	$985,300	Cost-effectiveness was not provided by the project sponsor.	
Adm/oper	$0	$0	$0		
Total	$788,240	$197,060	$985,300	Annualized Cost = $985K x 0.142 CRF = $139,870	
Total annualized public cost:	$146,900			C/E = $139,870 / (0.088 x 365) = $4,355/ton	
Annual revenues:	None				
Net public cost:	$985,300				
Annual private cost	NA				
Total net cost	$985,300				

NOTE: Assume that these facilities would be demanded all days of the year, consistent with Interstate trucking activity.

APPENDIX D. CMAQ PROJECT COST EFFECTIVENESS CALCULATIONS

Category	Subcategory	CMAQ PROJECT ID	PROJECT DESCRIPTION	Annual Emissions Reduction (kg/year)					Annualized Project Cost	Cost Effectiveness ($/ton)				
				VOC	CO	NOX	PM10	PM2.5		VOC	CO	NOX	PM10	PM2.5
Traffic Flow Improvements	Traffic Signalization	MI20020058	Ryan Rd. 8 Mile Rd. To 23 Mile Rd.	2,715.0	-16,858.9	-1,706.1	0.0	0.0	$110,256	$36,840	-$5,933	-$58,627	$0	$0
Traffic Flow Improvements	Traffic Signalization	LA20040001	Continuous Flow Intersection-Airline @Sherwood Forest											
Traffic Flow Improvements	Traffic Signalization	KY20050008	Fiber Optic Cable Installation	1,916.6	7,052.1	649.3	11.9	7.7	$904,773	$428,247	$116,390	$1,264,080	$68,974,198	$106,166,167
Traffic Flow Improvements	Traffic Signalization	OH20050033	West Main Street New Signals	31,311.2	115,206.8	10,607.7	194.4	126.3	$61,553	$1,783	$485	$5,264	$287,233	$442,113
Traffic Flow Improvements	Traffic Signalization	TN20050016	Sr-169 Cedar Bluff To College St	3,461.5	12,736.2	1,172.7	21.5	14.0	$98,414	$25,792	$7,010	$76,132	$4,154,132	$6,394,106
Traffic Flow Improvements	Traffic Signalization	KY20060009	Lane Use Control - Reversible Lanes	1,975.2	-14,400.0	-828.3	0.0	0.0	$5,078	$2,332	-$320	-$5,562	$0	$0
Traffic Flow Improvements	Traffic Signalization	Not yet assigned	2 Lane Roundabout At Fuller And Washington	304.7	1,121.0	103.2	1.9	1.2	$74,536	$221,937	$60,319	$655,103	$35,745,553	$55,020,116
Traffic Flow Improvements				113.8	601.5	86.7	0.0	0.0	$467,981	$3,730,970	$705,808	$4,896,080	$0	$0
Traffic Flow Improvements	Freeway Management	LA20030008	Baton Rouge Phase 2 Its	15,300.6	197,991.9	39,702.2	1,438.4	934.5	$443,109	$26,272	$2,030	$10,125	$279,456	$430,144
Traffic Flow Improvements	Freeway Management	WA20040027	Duwamish Its	2,973.6	16,223.8	2,268.0	0.0	0.0	$318,190	$97,073	$17,787	$127,274	$0	$0
Traffic Flow Improvements	Freeway Management	CT20050001	I/M System Design And Construction	4,173.7	15,563.5	1,322.4	0.0	1.5	$218,725	$47,541	$12,752	$150,051	$0	$135,906,590
Traffic Flow Improvements	Freeway Management	MI20050090	3 Changeable Message Boards On I-75	28,431.5	-78,068.8	-942.5	0.0	0.0	$23,082	$737	-$268	-$22,218	$0	$0
Traffic Flow Improvements	Freeway Management	Not yet assigned	Alabama Service And Assistance	8,102.6	29,812.7	2,745.0	50.3	32.7	$871,432	$97,568	$26,517	$287,995	$15,714,389	$24,187,834

Category	Strategy	Project ID	Patrol											
Traffic Flow Improvements	High-Occupancy Vehicle Lanes	TX20020069	Dallas How Interchange Ih 635 Us 75	1,351.7	18,950.6	632.2	0.0	0.0	$28,194,393	$18,922,053	$1,349,691	$40,458,088	$0	$0
Shared Ride Programs	Regional Ridesharing	MD20020010	Maryland Ridesharing Program	12,214.8	153,913.7	13,830.8	551.0	254.5	$1,200,211	$89,139	$7,074	$78,724	$1,976,153	$4,278,198
Shared Ride Programs	Regional Ridesharing	PA20050202	University Of Pitt Tdm	3,778.6	48,421.7	3,899.3	144.7	66.8	$358,669	$86,112	$6,720	$83,444	$2,248,489	$4,867,780
Shared Ride Programs	Regional Ridesharing	Not yet assigned	Rideshare Program	1,400.0	17,674.1	1,570.0	62.1	28.7	$762,503	$494,078	$39,138	$440,593	$11,136,120	$24,108,721
Shared Ride Programs	Vanpool Programs	UT20020006	Expansion Of Uta Vanpool Leasing Program (Salt Lake And Ogden)	733.9	8,871.4	725.2	30.3	14.0	$128,200	$158,471	$13,110	$160,365	$3,832,426	$8,321,080
Shared Ride Programs	Vanpool Programs	UT20050005	Expansion Of Uta Vanpool Leasing Program (Ogden/ Layton)	781.1	9,061.0	837.8	34.9	16.1	$47,676	$55,372	$4,773	$51,622	$1,238,280	$2,685,195
Shared Ride Programs	Vanpool Programs	KY20060004	New Passenger Vans For Lextran	820.9	10,440.1	944.4	40.0	18.5	$30,643	$33,865	$2,663	$29,437	$694,959	$1,505,981
Shared Ride Programs	Park-and-Ride Lots	MD200017	I-95/Md 279 And I-95/Md 272	2.2	48.4	3.8	0.1	0.1	$20,497	$8,453,705	$384,182	$4,908,851	$128,188,176	$277,516,130
Shared Ride Programs	Park-and-Ride Lots	WI20000034, WI20000035	Lake Geneva Park And Ride	464.2	5,764.4	565.0	23.6	10.9	$7,408	$14,478	$1,166	$11,894	$284,595	$616,124
Shared Ride Programs	Park-and-Ride Lots	MD20020001	Md 210 At Md 373 Peak And Ride Construction	563.5	10,071.3	852.9	34.0	15.7	$180,050	$289,871	$16,218	$191,501	$4,802,166	$10,396,268
Shared Ride Programs	Park-and-Ride Lots	KY20050012	Tank - Walton/Union Park And Ride Development	460.7	6,677.2	612.1	25.2	11.6	$143,695	$282,971	$19,523	$212,970	$5,175,501	$11,204,505
Shared Ride Programs	Park-and-Ride Lots	WA20010004; WA20050035	Mountlake Terrace Station	1,487.7	18,475.8	1,810.9	75.7	35.0	$1,741,954	$1,062,245	$85,532	$872,653	$20,880,023	$45,203,414

TDM	TDM	CO20010042	Coordinate Telework Program	6,683.7	85,192.1	7,112.0	270.5	125.0	$116,371	$15,795	$1,239	$14,844	$390,221	$844,795
TDM	TDM	DC20020006	Employer Outreach And Bicycles Fy 2003	117.5	1,547.5	101.4	3.2	1.5	$18,832	$145,450	$11,040	$168,487	$5,417,471	$11,728,348
TDM	TDM	DC20050008	Fy06 Guaranteed Ride Home	50,604.3	637,587.5	57,326.3	2,284.5	1,055.2	$1,941,109	$34,798	$2,762	$30,718	$770,832	$1,668,784
TDM	TDM	RI20050010	Ozone Alert Days	61.2	788.6	61.1	2.2	1.0	$194,284	$2,879,727	$223,494	$2,882,338	$79,844,584	$172,856,506
Bicycle Pedestrian	Bicycle Pedestrian	MA20020040	Swansea Old Warren Rd. Bikeway Facility	138.0	1,834.8	111.6	3.2	1.5	$167,471	$1,100,607	$82,804	$1,361,137	$47,568,807	$102,982,286
Bicycle Pedestrian	Bicycle Pedestrian	IN20050009	Bike Path To Pinhook Park	35.7	475.1	28.9	0.8	0.4	$237,332	$6,023,188	$453,217	$7,440,882	$259,655,935	$562,132,272
Bicycle Pedestrian	Bicycle Pedestrian	Not yet assigned	Bike Depot	126.2	1,504.1	104.3	3.1	1.4	$131,797	$947,239	$79,494	$1,146,475	$39,181,263	$84,823,991
Bicycle Pedestrian	Bicycle Pedestrian	Not yet assigned	NYC Cyclistnet	715.4	8,524.0	591.0	17.3	8.0	$434,833	$551,441	$46,278	$667,428	$22,809,613	$49,380,807
Transit Improvements	New Bus Services	WI20000004	City Of Racine New Sunday Bus Service	150.8	0.0	166.4	0.0	0.0	$250,886	$1,509,283	$0	$1,367,788	$0	$0
Transit Improvements	New Bus Services	NY20050028	S92 New Bus Route	951.2	13,525.2	229.6	20.1	-0.4	$135,907	$129,618	$9,116	$537,083	$6,139,286	-$275,107,412
Transit Improvements	New Bus Services	RI20050003	Service Initiatives Route 30 And Route 12	729.2	18,952.9	1,343.8	48.6	22.4	$328,780	$409,027	$15,737	$221,962	$6,137,442	$13,287,023
Transit Improvements	New Rail Services	UT20020001	Light Rail Vehicles	4,592.3	54,122.7	4,524.7	161.0	74.4	$443,012	$87,514	$7,426	$88,822	$2,495,576	$5,402,702
Transit Improvements	New Rail Services	TX20030147	TRE Double Tracking Of Segments	16,644.3	207,433.7	18,194.1	708.4	327.2	$7,631,105	$415,928	$33,378	$380,497	$9,772,506	$21,156,615
Transit Improvements	New Rail Services	CT20050027	Rail Station Platforms And Bridge	1,716.0	30,670.7	2,597.5	103.6	47.8	$261,293	$138,134	$7,729	$91,257	$2,288,406	$4,954,197
Transit Improvements	Service Upgrades/Amenities	MA20020069	Fitchburg ITC Parking Garage	2,201.5	39,347.5	3,332.4	132.9	61.4	$92,327	$38,046	$2,129	$25,135	$630,289	$1,364,520
Transit Improvements	Service Upgrades/Amenities	MO20040023	Operation Welcome Aboard Infrastructure	709.1	9,105.4	723.0	26.6	12.3	$190,914	$244,252	$19,021	$239,538	$6,519,661	$14,114,492

Category	Sub-category	ID	Project											
Transit Improvements	Service Upgrades/Amenities	NY20040006	Suffolk County Transit Marketing Program	666.4	6,564.8	595.1	19.0	8.8	$123,607	$168,276	$17,081	$188,430	$5,911,468	$12,797,808
Transit Improvements	Service Upgrades/Amenities	OH20050008	Laketran AVL-MDT System	327.4	4,220.6	326.2	11.7	5.4	$538,585	$1,492,350	$115,766	$1,497,977	$41,600,710	$90,061,880
Transit Improvements	Service Upgrades/Amenities	Not yet assigned	Commuter Rail And Utility Construction	1,162.9	20,785.1	1,760.3	70.2	32.4	$14,227	$11,098	$621	$7,332	$183,862	$398,046
Technology and Fuel Programs	Conventional Bus Replacements	MD20020008	100 Replacement Local Buses	5,610.0	0.0	62,337.0	0.0	0.0	$9,179,510	$1,484,405	$0	$133,589	$0	$0
Technology and Fuel Programs	Conventional Bus Replacements	Not yet assigned	Ham - Sorta - 61 Replacement Buses	2,506.1	3,026.1	9,237.8	0.0	0.0	$2,353,541	$851,946	$705,551	$231,126	$0	$0
Technology and Fuel Programs	Alternative Fuel Vehicles/Fueling Facilities	ME20020020	Fast Fill Compressed Natural Gas Facility	719.7	0.0	553.8	0.0	0.0	$192,912	$243,173	$0	$316,011	$0	$0
Technology and Fuel Programs	Alternative Fuel Vehicles/Fueling Facilities	PA20020062	Alternative Fuel Buses	750.0	3,000.0	22,750.0	0.0	0.0	$2,428,240	$2,937,150	$734,288	$96,829	$0	$0
Technology and Fuel Programs	Alternative Fuel Vehicles/Fueling Facilities	CT20050025	Ct Clean Fuels Program	1,032.8	0.0	1,911.0	0.0	0.0	$172,670	$151,676	$0	$81,971	$0	$0
Technology and Fuel Programs	Alternative Fuel Vehicles/Fueling Facilities	Not yet assigned	Purchase 3 Forty Foot Urban Transit Cng Buses	540.0	2,743.2	1,562.4	0.0	504.0	$375,688	$631,146	$124,241	$218,138	$0	$676,227
Dust Mitigation	Dust Mitigation	CA20040439	Graaf Avenue Paving Project	0.0	0.0	0.0	52,195.0	0.0	$20,817	$0	$0	$0	$362	$0
Dust Mitigation	Dust Mitigation	ID20040003	Lincoln Ave Sandpoint	0.0	0.0	0.0	45,633.1	0.0	$33,710	$0	$0	$0	$670	$0
Dust Mitigation	Dust Mitigation	ID20050017	Liquid De-Icer Truck	0.0	0.0	0.0	1,824,680.0	0.0	$29,865	$0	$0	$0	$15	$0
Freight/Intermodal	Freight/Intermodal	ME20000004	Rail Siding Construction	97.2	488.0	2,038.7	69.2	56.9	$41,096	$383,558	$76,393	$18,287	$538,502	$655,448
Freight/Intermodal	Freight/Intermodal	ME20020005	Rehab/Replace Tracks On Sprague Industrial Spur	24.9	148.3	830.6	21.5	17.7	$54,723	$1,997,735	$334,684	$59,767	$2,304,829	$2,805,363

Category	Subcategory	Project Code	Project Name											
Freight/Intermodal	Freight/Intermodal	PA20020059 PA20030090	Westmoreland Itc	2.2	250.8	4,374.3	86.2	70.8	$1,028,708	$424,194,830	$3,721,007	$213,345	$10,831,847	$13,184,175
Freight/Intermodal	Freight/Intermodal	NY20040036	Arlington Intermodal Yard	23,385.8	117,353.4	378,629.3	13,059.5	10,729.5	$949,286	$36,825	$7,338	$2,274	$65,942	$80,263
Freight/Intermodal	Freight/Intermodal	PA20040076	Norfolk Southern Rail Ext	1,058.5	5,256.4	19,944.8	690.0	566.9	$1,318,453	$1,130,019	$227,547	$59,970	$1,733,406	$2,109,846
Freight/Intermodal	Freight/Intermodal	CT20060022	Rail And Utility Construction	176.8	876.4	3,238.3	112.6	92.5	$174,141	$893,459	$180,259	$48,784	$1,402,472	$1,707,043
Engine Retrofit Technologies	Diesel Engine Retrofits	MD20010025	Bus Engine Upgrade	0.0	0.0	0.0	0.0	9,040.2	$5,094,530	$0	$0	$0	$0	$511,236
Engine Retrofit Technologies	Diesel Engine Retrofits	NY20040032	Wcdot Diesel Engine Retrofit Of 177 Transit Buses	1,059.7	16,446.5	0.0	0.0	823.4	$311,010	$266,254	$17,155	$0	$0	$342,657
Engine Retrofit Technologies	Diesel Engine Retrofits	PA20040011	Emissions Reduction Device	1,832.5	27,805.0	0.0	1,635.0	1,432.5	$371,869	$184,095	$12,133	$0	$206,333	$235,500
Engine Retrofit Technologies	Diesel Engine Retrofits	OR20050011	Exhaust After-Treatment Controls On Trash Collection Trucks	371.8	647.4	0.0	72.8	0.0	$12,457	$30,395	$17,456	$0	$155,229	$0
Engine Retrofit Technologies	Diesel Engine Retrofits	Not yet assigned	3 Locomotive Repowers	3,386.4	0.0	44,914.0	0.0	1,251.2	$1,042,808	$279,358	$0	$21,063	$0	$756,090
Engine Retrofit Technologies	Diesel Engine Retrofits	Not yet assigned	Orangetown Diesel Vehicle Retrofits	134.2	521.6	0.0	54.9	42.7	$100,116	$676,780	$174,142	$0	$1,654,351	$2,127,022
Engine Retrofit Technologies	Diesel Engine Retrofits	Not yet assigned	Diesel Engine Retrofits Of Rockland County Vehicles	42,800.7	295,801.2	0.0	42,196.8	38,393.4	$323,016	$6,847	$991	$0	$6,945	$7,632
Engine Retrofit Technologies	Truck Stop Electrification	TN20030011	Tse: Idleaire 100 Units Watt Rd.	0.0	0.0	49,275.0	1,343.2	1,343.2	$163,332	$0	$0	$3,007	$110,313	$110,313
Engine Retrofit Technologies	Truck Stop Electrification	KY20060013	50 Advance Travel Center Electrification (Idle Aire)	0.0	17,060.1	40,212.1	671.6	671.6	128,490.99	$0	$6,833	$2,899	$173,563	$173,563

Engine Retrofit Technologies	Truck Stop Electrification	TN2006026	Install Idleaire At Sites In Jefferson Co	0.0	0.0	29,072.3	792.1	792.1	$146,881	$0	$0	$4,583	$168,232	$168,232